AF589390

Urban Environment and Social Wellbeing in Second Class Indian Cities

The Author

Dr. Aboo Ishaque P.K. born in a village known as Mampad of Malappuram district of Kerala, India. He did graduation in Geography from HM College, Manjeri (Calicut University) and did his post graduation and Doctor of Philosophy from Aligarh Muslim University, Aligarh, U.P. in the same subject. He presented several research papers in International conferences and published several research papers in different journals and proceedings of international conferences.

He is a resource person to several Civil Service Academies in Malabar region and a life member of Global Lake Ecology Observation Network (GLEON), at present working as a Guest Lecturer in the Department of Geography, Kannur University, Kerala.

Urban Environment and Social Wellbeing in Second Class Indian Cities

–Author–
Dr. Aboo Ishaque

2015
Scholars World
A Division of
Astral International Pvt. Ltd.
New Delhi – 110 002

Cataloging in Publication Data—DK
Courtesy: D.K. Agencies (P) Ltd. <docinfo@dkagencies.com>

Ishaque, Aboo.
Urban environment and social wellbeing in second class Indian cities / author, Aboo Ishaque.
p. cm.
Includes bibliographical references (p.).
ISBN *9789390384686* (Hardbound)

1. Urban ecology (Sociology)—India—Calicut. 2. Well-being—India—Calicut. I. Title.

DDC 307.760954 23

Published by : **Scholars World**
A Division of
Astral International Pvt. Ltd.
– ISO 9001:2008 Certified Company –
4736/23, Ansari Road, Darya Ganj
New Delhi-110 002
Ph. 011-43549197, 23278134
E-mail: info@astralint.com
Website: www.astralint.com

FOREWORD

The social area analysis of urban places has a paramount importance in contemporary world because of the growing social pathologies and delinquencies associated with the rapid urbanization experienced during last few decades. The deteriorating physical environment of the urban areas due to the congestion and overcrowding is also a subject of great concern. Reduction of the disparity between the slums and elite class residential areas is the important task of the urban policy makers and planners. The urban environment and social wellbeing in second class Indian cities is an appreciable attempt from the author to delineate the social and environmental segments within an urban area.

The author tried to outline the concepts of urban environment and social wellbeing and traced the development of the concerned studies and researches in detail. This book presents a practical model of the urban social and environmental area delineation by giving the example of the city of Calicut. It is very useful to those who are interested to understand the great contrast between the lives of economically, socially and politically downtrodden peoples and others. The resource allocation mechanism must be rearranged or organized according to the areal variations to reduce and finally to eradicate the inequalities in concerned administrative limits.

This book is a handy source to those interested to know about the physical and cultural landscape of Kerala in general and of the Calicut city in particular. The fieldwork carried out for this study and sample size both indicate efforts and time spend. It is highly appreciable as most geographical analyses are armchair studies based on secondary sources. Fieldwork is a very important area in Geography and departments of geography must train students in conducting fieldwork. It increases value of the present book.

As a whole, the book gives a good insight in to the urban environmental and the social wellbeing aspects of the second class Indian cities and will be found useful for the researchers, academicians and planners in the field.

Prof. Fakhruddin

Department of Geography

A.M.U Aligarh, U.P

PREFACE

The number of urban dwellers and level of urbanization are growing at an accelerated rate as a response to the industrialization and associated economic, social, cultural and technological development. At present, more than a half of the world's population lives in urban places. The quality of the urban environment as a living space for the inhabitants, therefore, is an issue of fundamental concern for researchers, policy makers and individuals. A strong interest in the measurement of social wellbeing has developed over the last few decades in order to know the quality of life as it is the principal objective in the course of social development.

The present study has been undertaken to outline the urban environment and the social wellbeing of the coastal city of Calicut in Kerala. It is a medium sized city in India with its unique demographic peculiarities viz. high literacy rate, low fertility rate and negative population growth rate. The main objectives of the study are to assess the important factors which control the residents' wellbeing and to find out the relationship between the urban environment and the social wellbeing of the city. The present study is principally based on primary data which is collected through extensive household level sample survey.

I take this opportunity to express my deep gratitude to Prof. Fakhruddin for giving valuable time to check the manuscript and guidance to complete this work with enthusiasm. I also feel obligation to all faculty members, my colleagues and non-teaching staff of the Department of Geography, Aligarh Muslim University Aligarh, for providing necessary facilities and useful suggestions.

Finally I present my appreciation to my beloved mother who escort me to the world of knowledge, my Little Faizan for his cheering attitudes, my life-mate Navar for her support and patience, my siblings Mujeeb Rahman, Sajeena and Rubeena with their spouses Cheriyon, Abi and Inni for their moral and financial support. Last but not least thanking to little blossoms of our home Minnu, Neenu, Monu, Shalu, Minu, Richu, Hani and Nainu along with my younger brother Riyas.

Aboo Ishaque P.K

CONTENTS

LIST OF TABLES

LIST OF FIGURES

1

INTRODUCTION

Visualization of the interaction between the human being and the natural environment is the heart of geographical investigations. The mode of human interaction has undergone a rapid twist along with the urbanization from the deterministic and sustainable way of life to somewhat possibilistic urban way of life. As a result the urbanization and associated phenomenon became one of the burning topics of interest among many disciplines like Geography, Sociology, Economics and Demography, finally an interesting multidisciplinary subject i.e. urban social geography has evolved.

Urbanization preceded by industrialization led to rapid demographic growth rather than of the urban attraction of the urban way of life, the inhabitants of rural districts are flooded in to towns and cities. The immigrant populations create wide belts of miserable slums while the host town or city lacks the resources to provide the newcomers with jobs, decent housing, minimum public services, with even the rudimentary protection for health and hygiene (Reissman 1964, p.161). A strong interest in the measurement of social well being has developed over the last few decades. There has been an increasing awareness on the part of the public and policy-makers of the wide-ranging effects of problems such as pollution, crime, lack of housing and sanitation etc. as they affect the total spectrum of the society. It is realized that material and economic development alone cannot eradicate the above problems (Leslie et al. 1978, p. 457)

Enhancement of quality of life is always amongst the principal objectives in the course of social development. The indicators have been widely employed as measures of people's wellbeing, its achieved level, and changes over time. Researchers are not satisfied with works that assess the quality of human life and monitor the evolution of the societal environment in monetary terms alone, and they do not intend to surrender conventional quantitative computation methods. Neither universally accepted definition of 'quality of life', nor methods of its measurements have yet been established. So the research on quality of life in most cases comes across two long-standing issues.

First, there is no standard way of establishing the conceptualization of 'quality of life' which tends to the level of achieved satisfaction in both the basic materialistic needs and emotional needs. Quality of life research, therefore, is concerned with an extensive range of topics, such as individual physical and mental health, wellbeing, satisfaction, family, work, housing, social relations, political and cultural lives, social ethics. Consequently, researchers have never arrived at any compromise in the selection and weighting of indices in empirical quality of life studies. Second, views on how 'quality' should be assessed. Quality implies value judgments and ranking; so quality of life might be ranked from high to low, and from positive to getting worse. Hence, comparisons between communities or periods often show difficult (Chan et al. 2005, pp. 260-261)

Urban geography has moved a considerable way in the last five decades. In the beginning it was predominantly based upon morphological views of urban structure with some expanding interest in economic functions of towns but very little awareness of the social qualities of urban life and organization. The spatial analysis paradigm has at all time created problem for human geographers. It was the mere complexity of technical procedure for some, and was danger of an increasing vagueness of reality and apparent stereotyped even much quantitative analysis for others (Herbert and Smith 1979, pp. 2-3)

Social problems are clearly both culturally and historically specific and they reflect the prevailing values, ideology, structure of the existing social formation, and the nature of the urban environment and the way in which the people perceive conditions of life within them. Social problems also will relate to the level of development of the economy, its capacity to support social services, the degree of cohesion or conflict within the population, the occurrences of forces that threat the existing social order, the strength of social control, and so on. Until late 1970s the collection of data on social problems conditions kept on a highly unsystematic method. Statistics on the incidence of crime, delinquency, illness, various forms of 'social deviance' etc. were generated largely as by-products of the administrative procedures (Herbert and Smith 1979, pp.13-14)

The geographical variations in social conditions are vital to determine the social state of the nation. Aggregate national statistics are a summation of the conditions of people distinguished by location of residence as well as in other respects, therefore, they fail to show the local situation, and hide the extreme. There is shortage of adequate information on areal variation in social conditions. It is due to the failure to recognize the importance of the geographical component of national social problems, which fallen between the two academic schools of geography and sociology. Geographers have usually ignored contemporary social problems, and sociologists, generally given little care to areal variations in the incidence of social condition (Smith 1973, pp. 4-7)

The various national level indicators of wellbeing like Gross national product (GNP) or crime rate make difficult to argue convincingly that these indicators alone provide an adequate measure of the wellbeing of the population. This type of dilemma led to the broadly applicable series of social welfare using social indicators for the assessment of the social wellbeing in community level, using the functional areas of public safety, health, education, community service etc. (Zenher 1997, pp. 2-3)

Factor analysis and principal component analysis have been used prominently in geographical research or social area analysis from 1970s. These statistical techniques are generally used to extract the underlying dimensions of variance within large matrices of numerical data, by deriving a relatively small number of composite variables -factors- accounting for a relatively large share of the original information. This is possible because many individual variables are quite highly inter-correlated, thus differentiating between areas in a similar manner. The territorial units of observation can be given scores on these composite variables (Smith 1973, pp. 11-12).

The differences in the structure of economic production and the pattern of population and settlement are usually accompanied by pronounced spatial disparities in wellbeing because of variations in access to basic goods, services, and amenities (Hall 1984). Quality of life implies a rather personalized concept whereas wellbeing refers to aggregates of people defined by area of residence more appropriately addresses the welfare of some social groups (Smith 1973, p. 66). Compared with the way in which national economic accounts are kept, the recording of the social state of nations was unsophisticated in the extreme. Mounting concern at this situation, coupled with shifting societal pre-occupations in the direction of more social aspects of life, came to a head in the latter part of the 1960s in the 'social indicators movement'(Herbert and Smith 1979, p. 14).

The basic tasks which can define the scope of human geography are descriptions, explanations, evaluation, prescription and the implementation. Description involves the empirical identification of territorial levels of human wellbeing, or human condition. Explanation covers the how of the description. It involves identifying the cause-and-effect links among the various activities undertaken in society, as they contribute to determine who gets what where. Evaluation involves making judgements on the desirability of alternative geographical states, and the societal structures from which they arise. Prescription requires the specification of alternative geographical states, and alternative societal structures designed to produce them. This is the process of planning the spatial organization of human activity, i.e. spatial reorganization. Prescription involves answering the ethical question of who should get what where. Implementation is the final process of replacing a state deemed undesirable by something superior (Smith 1979, pp. 9-10).

1.1 Case Study

This study is an endeavour to outline the urban environment and the social wellbeing of the various residential areas of Calicut city to present a practical model of social area analysis. For this study the units of residential area are the wards. Wards are the administrative divisions of the city based on the geographic realities and so exhibit more or less homogeneous character. So the wards are the meaningful units of the area for analysis of wellbeing or quality of life in city or micro-level studies. The author has investigated the urban environment and the social wellbeing of each of the wards with the help of certain selected indicators.

The area of the study is the Calicut city, Kerala as demarcated by the Municipal Corporation and it consists the actual city and the suburbs. It is a medium sized coastal city and is the third largest and one of the main commercial cities of Kerala. The city was founded on a marshy tract along the Arabian coast in 1034 A.D. Calicut has been a multiethnic and multi-religious town since the early medieval period. Hindus form the largest religious group, followed by Muslims and Christians.

Many works already done on the quality of life at city level in India and abroad, but the city level enquiries in terms of wellbeing, for medium sized cities, using sound methodologies are not as much in the country. This study therefore is an effort to outline the wellbeing in a medium sized city and the pattern of its distribution. The research problem is very important because it undertake the study of an aspect which affects millions of people who are living in urban areas all over the India especially southern India. The trend of urbanization and the moving of people from rural to urban areas became a matter of increasing importance. So the information about the environment and wellbeing to be collected and analyzed to sustain and improve the habitat in which large number of people are living. It is this realization that stimulated the author to undertake the study of urban environment and social wellbeing of Calicut city.

Although the spatial pattern of the environment and the social wellbeing of the Indian cities are highly varying, there are considerable similarities between them especially in the case of south Indian cities. It can say that Calicut city is a representative of South Indian cities. The city of Calicut has been selected because it is a historical, cultural and medium sized city, and moreover is familiar to the author.

The living condition and extends of urban problems exclusively for developing countries are portrayed by Mc Gee: the rapid growth of the third world cities is evident everywhere in their physical appearance… housing and city population are forced in to squatter settlement of flimsy miserable huts constructed of makeshift materials which occupy any vacant land in the interstices or fringes of the city… The attempt to build up community and civic pride breaks down under the impact of population growth and "shared poverty" in to a condition of urban anarchy. The end product is that the cities of the third world become wastelands, conglomerations of millions of individuals… (McGee 1977, pp.18-19)

The distinction of residential areas over city space in Indian cities goes hand in hand with community or ethnic differentiation of the city space. Access to resources and power in these areas has been largely determined by the ethnic status of people, that is the position of population groups in the hierarchy of caste, or race or religion determines their socioeconomic status and political power. In the specific example of India, access of a community to power and therefore level of its influence on the geographical allocation of scarce public resources depends on its position with regards to dominance and caste hierarchy (Fakhruddin 1991, p.4).

The analysis of the social wellbeing which depends on the quality of urban environment in the context of residential areas falls in the domain of 'urban ecology'. This approach still recently had been employed to inquire the socioeconomic structure of urban populations and its residential distribution within the urban areas. These analyses are based on the matrices which contain information regarding the socioeconomic characteristics of a population, housing conditions, amenities, and facilities, security feeling and such relevant variables of urban life in small areas within the city (Fakhruddin 1991, pp. 5-6).

In India there is a shortage of relevant classification of cities having population more than one lakh. But pay commission of central government of India classified Indian cities to X, Y and Z categories are more commonly known as Tier-I, Tier-II and Tier-III cities respectively. Six mega cities of India viz. Mumbai, Delhi, Bangalore, Chennai, Hyderabad and Kolkata are grouped as tier-I and the 67 cities which include Calicut is considered as tier -II and all other cities as tier-III. In this sense, the author used the term second class cities for the cities of medium population size.

1.2 Data and Sampling

The selection of data is basically influenced by the concept of social wellbeing developed by the school of social geography since 1960s. A large number of variables containing information with regards to population characteristics and condition of life are used in this analysis, since there is no single comprehensive source of required information. Even the census of India can't give population statistics on the city sub-areal level of wards. The information available in the census table relates to population size, sex composition, scheduled caste and scheduled tribes, literate population, number of households, work participation rate and number of occupied residential houses. The shortage of required data from census table compelled to explore other sources like government and quasi- government offices and agencies which collect and maintain statistics on population and other aspects of city life.

Secondary data available in different places in different forms are not sufficient to draw out the social wellbeing and the urban environment, therefore one and half (1.5) per cent sample survey of total household in the city is conducted. The unit of survey was households as the purpose of study was to analyse household characteristics as well as the condition of housing and immediate environments in which they live. The sampling procedure adopted can be described as stratified random, the stratum selected as the wards. The care is taken in the survey to represent the households of different categories, communities and localities present in the wards. For this purpose the author interviewed the old shop keepers, political representatives and NGO field workers etc. and collected the information and location of particular phenomenon on the respective ward.

The city of Calicut is divided in to fifty-one wards and 80528 households according to 2001 census. From each wards one and half percent of sample is selected randomly which have a range of between twenty and thirty households according to the total number of households in respective wards. The households of selected houses were asked questions related the wellbeing through a prepared schedule (appendix C), their responses and some observation of the surveyor were recorded. The information collected was arranged at ward level for analysis

The variables were chosen as indicators of the urban environment and social wellbeing and they related to housing structure, amenities and facilities, demography and ethnicity, education and income and the security and migration. The absolute values are not so fit to comparison in ward level, so they are transferred to percentages and ratios. The present study is based on both primary and secondary data. The secondary sources include Census Reports, economic and statistics report and different records of Corporation office.

1.3 Techniques of Analysis

The survey and observation constitute a large number of important variables which explain the urban environment and the social wellbeing of Calicut city. The representation and analysis of this much of data was a great task to the author. Several simple methods not allow the multivariate structure without which the study cannot be complete. Appropriate operational techniques for this task are found in two research traditions in geography: 1) multivariate regionalisation and 2) factorial ecology. The method of multivariate regionalisation has developed and spread rapidly after the publication of Ginsburg's atlas of economic development. Many of the earlier attempts have employed simple additive techniques involving ranking and classification of indicators according to some theoretically determent criteria. Later this methodology was modified under the 'social indicators' approach that reacted sharply to the over emphasis on economic criteria as the measure of human wellbeing. As a result, more and more social indicators have been incorporated in the regional analysis of the development. Since the relationship among these varied indicators of development have become uncertain, procedures of standardisation have been adopted so that the transformation of indicators may entail there addition into various categories of the development.

Methodology of factorial ecology developed in the early 1960s has grown out of an older tradition of social area analysis in urban geography. Social Area Analysis is the technique for identifying segments within a city, or it is the process of identifying the "urban sub-communities". The basic unit for social area analysis is the census tract, which is a group of contiguous blocks in a city, contains a definite number of people and intended to be a socially homogeneous as possible (Reissman 1964, p. 86). Factorial ecology employs a variety of mathematically rigorous methods of factor analysis to reduce a large number of socio economic and environmental indicators into a few underlying dimensions. Unlike the methodology of multivariate regionalisation and social area analysis with structure variables according to some theoretical constructs, it allows the construct to emerge from the interrelations of the original variables from which a similar set of smaller number of variables is derived that reproduce the original relationships except for the restriction that derived variables are independent of each other. By combining standardised original variables and their loadings on computed variables (factors), original variables may be aggregated to exhibit regional distribution of new variables.

The methods of classification of variables in to major dimensions in the two traditions have their relative advantages. The additive method involved simple calculation and there is little ambiguity involved as all the subjective elements are usually known and made explicit. Moreover since they imply no assumptions of orthogonality of categories or dimensions, relationship among them may be evaluated and analysed. Such methods of classification are quite valid, if theoretical constructs are acceptable and the addition of the variables is also legitimate. However the assignment of equal rank-difference to varying magnitudes of a variable results in considerable loss of information. Standardisation procedure, usually that of normalisation of distribution overcomes much of the loss of information, nevertheless simple addition without giving consideration to the significance of the constituent indicators of a category can't represent a major part of reality.

The aforesaid problem is largely solved by the factor analysis method because the loading of variables on a factor (category) is their weight which is derived from their factual interrelationships. But the factor analysis procedure starts with a solution which is not mathematically unique. Therefore, there is no assurance that factors obtained would conform to the theoretically relevant or most important aspects of the reality. Generally apart from the first factor that is understood as an overall index, all other factors remain un-interpretable. Hence factors are subjected to relation to some theoretical criteria to make the factor structure more interpretable. But at this stage the problem become more complex. Though factor loadings remain orthogonal, factor scores do not. Thus at the final stage of this analysis ambiguity is involved in the interpretation of regional patterns of various dimensions.

The data from different sources is processed and complied at ward level which is taken as the unit of analysis. The processing has involved conversion of raw data into derived values as percentages, ratios, densities etc. In some cases weighted indices are also calculated which are explained at appropriate places. The appropriate operational technique used for this task is factor analysis. It can help to explore the way in which household differentiated over the city space. However some simple and some complex techniques are also used to classify wards and indicators into groups like Standard deviation method and Principal components analysis to reduce large number of variables in to smaller number.

Computation for this analysis has been carried on the computer present in the university computer centre of Aligarh Muslim University Aligarh using SPSS 11.0 version and indigenously developed SSP programme in FORTRAN. Besides these methods, advanced cartographic techniques and GIS-Arc View 3.1 programmes have been adopted.

1.4 Factor Analysis

Factor analysis is a statistical method used to describe variability among observed, correlated variables in terms of a potentially lower number of unobserved, uncorrelated variables called factors or it is a statistical technique, the aim of which is to simplify a complex data set by representing the set of variables in terms of a smaller number of underlying variables, known as factors. The technique is a branch of multivariate analysis and may also be described as unsupervised learning and an exercise in modelling. Steps for performing factor analysis are given below.

1. Collection of required data.
2. Arrangement of a variance-covariance matrix of the observed variables.
3. Selection of suitable number of factors for the particular problem.
4. Extraction of the initial set of factors.
5. Perform the factor rotation to a terminal solution.
6. Interpretation of the factor structure.
7. Construction of factor scores to use in further analyses

The factor analysis include Forty-nine variable relating to the urban environment and the social wellbeing as described by housing structure, amenities and facilities, demography and ethnicity, education and income, and the security feeling and migration of the inhabitants of the fifty-one wards of Calicut city. Computations for this analysis were carried on the system available in the computer centre of Aligarh Muslim University. The programme of the factor analysis available in the stored *Standard Subroutine Package* gives a principal components solution. The steps of computation are summarized below:

- Computation started with the transformation of original data matrix D for *n* observations on *m* variables in to a standard score matrix Z of n ×m order
- From the Z matrix an m × m order correlation matrix R is calculated which contain the product moment correlation coefficients.
- This correlation matrix was resolved in to a factor matrix *A* of *m*×*r* where r was the number of factor extracted. The programme employed can extract as many principal components as the number of variables, therefore, in the first instance all the factors were extracted. A histogram of the cumulative percentages of the variance explained by the successive factors and cumulative number of factors was constructed. By inspecting the rate of change in the explanation of variation by factors, the number of factors to be retained was determined.
- Since the original variables retained were not readily interpretable, the factor loading matrix A was rotated according to the normal varimax criterion to reproduce a new factor loading matrix. The criterion employed rotated the factor matrix to such a position where a minimum possible number of variables loaded high on each factor. The factor structure thus became simpler and more easily interpretable.
- From the matrix multiplication of the standard score matrix of n×m order and rotated factor matrix A of m×r order, a factor score matrix F of n×r order was obtained. The factor scores were then normalized to zero mean and unit

variance. These factor scores provided a measure of position of each ward on the new factor.

Here it should be pointed out that a factor analysis including variables of urban environment and social wellbeing might have a better exercise. This might have helped in the analysis of the intricate relationship between the two sets of variables. But the limitation imposed upon by the programme used, which is the number of variable should not exceed the number of observation, has not only restricted the analysis but also imposed a limited number of variables in the analysis. So many other important variables had to be omitted. But the relationship between the urban environment and the social wellbeing were qualitatively evaluated.

2

OUTLINE OF CONCEPTUAL DEVELOPMENT IN URBAN ENVIRONMENT AND SOCIAL WELLBEING

The review of relevant literature forms a prominent foundation in any academic work as it provides the background to and justification for the research undertaken. Study of the conceptual development through the time is a critical look at the existing works that is significant to the work that carrying out. The summarization of relevant research is also vital that it evaluate the works, show the relationships between different works, what are the methodologies have been incorporated in related works and also show how it relates to ongoing work.

The author would like to organize the conceptual development chronologically because the developments over time are crucial to explain the context of the urban environment and social wellbeing. The chronological system will be an effective way to organize the work. This is not a time to provide a summary of all the published work that relates to urban environment, but a survey of most relevant and significant works which- include books, edited books, research papers, articles and reviews- is done here.

Urban geography was greatly concerned with form in 1960s and 70s, and then the classical relationship of urban land use and morphology with emphasis on site, location, and physical conditions retrocede into the focus. Geographical concern began to consider the internal features of the structure of an urban system, and these features which are powerful among the environmental forces conditioning urban life, belong to the 'built environment' of the place and the socioeconomic characteristics of the population. New innovation in the geographical investigations such as Quality of urban environment came after 1970 when concern with human habitat became subject of more focussed research within the older traditions of 'quality of life', 'social indicators', 'human development' etc. as a dissatisfaction from gross economic measures of production and consumption as true indicants of human wellbeing and it developed as a new approach to the human wellbeing (Harrison and Gibson, 1976, pp. 14-19)

Urban environment and social wellbeing introduced into the sociological and geographical studies at an increasing rate. Many scholars addressed the problem of the social wellbeing. In this chapter, the author presents an analytical framework for urban environment and social wellbeing given by several experts.

In early sixties the treatment of the census data and selection of senses tracts were the foremost consideration in the urban and sociological studies which later associated with the social wellbeing. A new approach to the application of statistics to one aspects of urban life is demonstrated by Beshers (1962). By carefully limiting the analysis to the correlation of residential proximity, marriage, social caste or class structures, and occupational status, the author clearly demonstrates how census tract data, official registration of residences on marriage licences, and other information can be used to reinforce and corroborate sociological hypotheses. Ultimately he developed a theoretical model, empirically oriented, of the relationship between spatial distribution and social distance in the average American cities.

The crime and crime rate are well connected with the wellbeing and it was a sociological debate in sixties as a psychological function of the life. Osborn (1968) brought to light the debates by several well known social activists and criminologists about the crime in soviet region on which he referred Karpets who advocated crime as a social phenomena and temporary or accidental situation of material want. A similar scheme by Sakhrovis also mentioned which stresses the personality of criminals. Another group is the Psychologists who support individual factors in crime. And he also familiarizes the work of Chest i.e. 'A Difficult Book' which indicates the environmental influence in the crime.

The collaboration of urban geographers, urban planners and sociologists came in existence which intended for tackling the conundrum of expeditiously growing urbanization and associated phenomenon. A fantastic description of urbanization on developing countries is edited by Weitz (1973) organizing the papers and discussion of the sixth Rehvot conference. The peculiarity of the work is the organization of the different views of experts from Urban planning, Geography, sociology etc. In his book the author arranged the scenario of urbanization and its social, economic aspects and the policy, prospects and strategy of planning. The problems of environment, housing, services, financial and uncontrolled growth are given emphasis. The challengers are outlined and the same time the methods and means to overcome the same are discussed in details in the last parts especially to the slums and associated problems.

In 1970s the social wellbeing emerged as a social awareness especially among the urban geographers of United States and diffused it all over the world. This development of debates boosted the further detailed studies in the problems of city and began to provide the city and society a dimension of the social area and the methodology of social area analysis is emerged.

Smith (1973) managed to outline the territorial social indicators which led to the social indicators movements in America and then to whole the world. The work incorporated a wide range of conditions relating to the social wellbeing of the society and the quality of life of individuals. The use of multivariate regionalization or the factorial ecology is the most important contribution of this work to the urban social geography and urban planning. The work was a deviation from the merely economic based approach to monitoring the social wellbeing. Three distinct studies of intercity intra city and interstate levels are included in this study. The territorial social indicator then became an intense concern in the city and regional planning.

The further association of the economists, planners, geographers and sociologists give birth to an established discipline of urban social geography which deals with the demographic and environmental problems of city life. Robson (1975) out lined the social areas on urban centres along with fertile summery of vast literature and new approaches. He argued that the geographers must be give emphasis to economic and sociological approaches than the simple geometric pattern of the urban areas. The book portrays the mutual interaction of housing space, social space and physical space. As a shift from the previous studies, the author has given importance to the supply side than demand side of housing areas. And he also tried to identify the social areas within the city space.

On contemporary to Robson Knox from the geography side conceptualized the wellbeing and amalgamated the social indicators to monitor it. No one like Knox (1975) did conceptualize the social wellbeing and quality of life. He defined the level of living and wellbeing along with the tabulation of social indicators that in his opinion should have the quality of comprehensive, available at time series, geographical conciseness, and relation to public goal. According to him there should be some commonly accepted social indicator devoid of these the comprehensive measure of wellbeing will be collectively ineffective even though it has some individual significance or use. He simply explained how will measure the wellbeing or quality of life by giving a practical example from England and Wales with the usage of 12 constituents 20 aspects and 53 indicators.

The usage of the social indicators made better the appraisal of social area analysis, wellbeing and level of living correspondingly in India. An attempt to measure inter-state disparities in level of living in India is made by Ganguli and Gupta (1976) taking fifteen major states of the country for three period of time 1955, 1960 and 1965. The level of living has been analysed by means of a set of indicators classified in to eight components: nutrition, housing, medical care, education, clothing, leisure, security and environment. The first five indicators have been termed 'primary' and the last three are 'secondary' components. To show the overall index of level of living, both primary and secondary aspects are integrated.

The dimension of the social area analysis is further became a controversy. Coates et al. (1976) in their work conceptualized quality of life, wellbeing, and need and the problems involved in measuring the quality of life. Existing inequalities between areas are described at three levels: the international, the intra-national and the intra urban. The pattern emerged in the study is explained in three broad categories such as, the division of labours, accessibility and the territorial division. The authors put forward some policies for the removal of inequalities in social wellbeing.

A large number of works are overwhelmed to conceptualize the quality of life and wellbeing in different perspectives. The interaction of the experts from biological sciences drawn the subject in to other dimension, but soon they became a different medical based quality of life which emphasis only in individuals. Harrison and Gibson (1976) developed a book from a series of meeting held by the Royal Society Study Group on human biology in urban environments, and it reviews the available knowledge about the effects of urbanization in the developed world. The interdisciplinary figures are contributed in their work especially biological, medical science, sociology and urban geography experts. According to the authors the city has become a fact of life and for many a very uncomfortable fact of life. The environmental conditions, both physical and social, which now prevail in urban life, are very different from those experienced by man throughout the history of his evolution. Therefore one can expect signs of biological, psychological, and sociological problems in urban populations.

The advancement of the studies led to the indicators of quality of life itself turn out to be a debatable puzzle, that should it use separately or combine. Day and Weitz (1977) attempted to bring out several solutions to the problem of combining the individual indicators into overall comparative measures. They obtained promising results by grouping cities according to their similarity of their profiles on the social indicators. They used 14 indicators of quality of life for 18 largest metropolitan areas in the United States and classified them.

The quality of data is became a consideration in the field of social area analysis. Some advocates it should be primary in nature and the geographers must be in touch with the society to identify the problem. Smith (1977) again came with a monumental work in the field of human geography with a special thrust to wellbeing approach. The author organized the whole study in to three distinct bases of theoretical perspective, approach to practice and the case studies from South Africa and USA. In this study he advocates that the geographers have to do numerous things in the social development and for this purpose they must be field oriented and their research should be problem solving. He conceptualized the wellbeing at different scale and given a developed social indicator frame for the analysis of social indicators.

Now the problems of urban area are differing according to the nature and the process of genesis and the evolution of the towns because it has a considerable influence in the social, cultural and ethnical organization of the city dwellers which will control the wellbeing and quality of life.

An in-depth study of urban institutions and their cultural settings in many different societies and times is depicted by Fox (1977) in his book urban anthropology. The author explained the regal-ritual cities, administrative cities, mercantile cities, colonial cities and industrial cities in different perspective like urban adaptation, urban organization and urban ideology. Several examples of different cities and dynasties from all over the globe are cited for the easy understanding of the concepts.

The nature of wellbeing or quality of life indicators became a point of discussion as it should be subjective or objective. Wan and Barbara (1978) tried to visualize subjective wellbeing by using the methodology of automatic interaction detector analysis (A.I.D) for socio demographic variables and multiple classification analysis (MCA) for physical and psychological measures. The study reveals the relative importance of certain socio-demographic, physical and psychological factors in explaining variation in general wellbeing with respect to the age differences.

It is very clear that the environment especially built up environment has a direct and solid influence in the quality of life of individual and the wellbeing of society. Lapping (1979) commended over social theory of built-up environment by keeping the central theme of built-up environment as the location, arrangement and the relationship among people and things. Author recites the components of built environment from Wright as fresh food, independence of peoples employment, community structure, industry, business offices, educational facilities, housing patterns etc.

Gender discrimination and the attitude to the women in a society to a great extend associate with the overall wellbeing of them. Lal (1979) made an attempt to analyze the status of women in the urban family of Patna using the primary data collected by the method of quota sampling. Author conducted the study by choosing 10 tasks that put under three important behavioural field of the family. They are economic activities, child care and control and household duties. The main finding was that in this region there is a strict division of labours in household duties.

The ethnic and religion component are also have influence in the wellbeing of community level as the perceptions of quality of life differ from community to community. Pacione (1980) did a monumental work to differential quality of life in metropolitan village. The research identifies two major community groups along a number of social and behavioural dimensions and delimits the territorial extend of each with in the village space. Inter group differences in life style are also found to be related to the residents' assessment of their individual quality of life. A model of life satisfaction is also included to assess the actual components of good life for each of the communities in the study area. A demonstration of practical merits of a subjective view point for social planning also included in the work.

The discipline is begun to explore all attributes of social life that can affect the quality of life. Family life and satisfaction is an important factor in the social wellbeing. Harvey et al. (1981) examined the family adjustment and satisfaction in a rapidly developing resource community with primary data collected by sampling. Authors analyze community satisfaction, social life satisfaction, pattern of family interaction, parent child relationship, social support to children and teen agers, all are associated with the family satisfaction which is a distinct indicator of quality of family life. They find out that the well satisfied community is more stable and attracting migrants whereas dissatisfied families move out from the community.

The personal life and the conflicts are not independent components of city life. For a better society the mentality of helpfulness is compulsory among the peoples. Fischer (1981) tried to criticise the concept that the city life is associated with less public helpfulness and more conflicts. Author challenged the concept by evidences from the residence of urban area of California in terms of helpfulness, conflict, personal relation and psychological states.

The external urban environment is also contributing to the wellbeing that the society is crazy to avail the public facilities and amenities. The discipline began to incorporate the physical environment and the collaboration between them is evolved. Knox (1982) gives an introduction to the social pattern and process of urban areas which integrate the key themes of contemporary Urban Geography from both radical and traditional perspectives as a powerful interpretation of the social structure and functioning of the urban environment. It also includes the role of evolving transport technology in shaping urban forms, the social dimension of life in cities, economic and social structure of the cities and the location conflicts within the city.

Simultaneously Beyless M and Beyless S (1982) examined the quality of life indices employed in the measurements of wellbeing. Authors discuss the theoretical basis for the formation of quality of life, processes of selecting variables, and evaluated the methodology that has been employed in aggregating variables. The work analyses the aggregate social wellbeing that is the sum of the wellbeing of many individuals, given by Smith and the general welfare that is individual quality of life combining the psychological and physical inputs coined by Liu. In addition, the article gives a plan of methodology to quality of life construction.

Sinden (1982) tried to present a new application and evaluation of the Liu's methods of estimating comparative value of quality of life. The study covered 26 towns in New South Wales using the census data of 1971 and used 11 variables under five categories i.e. income, health, social order, housing and environment. The peculiarity of the study is the use of multiple procedures like modified exponential sampling, cube root standard deviation and proportional procedures. The main contribution of the work is that the value and individual component index of the quality of life.

The impact of urbanization and urban way of life on social and psychological relationship, and the contact among individuals and families are also very important. Palisi and Canning (1983) tested how urbanism affects social and psychological wellbeing using primary data of London, Los Angeles and Sydney. Path Analysis is used by considering visiting friends, visiting kin and marriage companion ship. It is found that deterministic theories can be rejected because more urban people have social relation not so much less than that of rural. They tested sub cultural theory mixed support and composition list. According to them best predictor of wellbeing is the extent of peoples' relationship.

The demographic aspects are also been included in the monitoring of the quality of life and social wellbeing. Andrews (1983) relating demographers' attributes of various population characteristics (size, growth, density, age/sex structures, migration, etc.) for the measurement of social wellbeing to provide new knowledge about the linkage of population and wellbeing that can enhance decision making about important population issues. With the reference of wide literature the author discusses the concept of social indicators and their relevance to population issues, social indicator movement and the dynamics of subjective evaluation with the current state of knowledge base.

Most of the third-world countries are characterized by marked socioeconomic regional variation. The differences in the structure of economic production and the pattern of population and settlement are usually accompanied by pronounced spatial disparities in wellbeing because of variations in access to basic goods, services, and amenities. Such variation in a typical third world country like Costa Rica is studied by Hall (1984). Five categories of indicators of socio economic wellbeing are used such as economic index, housing and amenities, education, health and security, and communication and societal participation. The author used the technique of factor analysis for the representation of spatial variation.

Herbert and Johnston (1984) opened an account to show the evolution of urban geography while they introducing the series of geography and the urban environment. They portrayed the twisting of the study from the 'site and situation to the inanimate environmental quality of city and its settings from 1920s 1960s and then to the quantification and behavioural alternatives. They plotted the present interest of the social scientists and policy makers out of physical scientists to eradicated inequality and injustice. They pointed out two important changes in the wider social environment that is the growing ecological consciousness of society and social and territorial justice.

Social and military conflict in the world left its impact on the world as many of the people shifted to refugee camps and the problems of wellbeing among them is get a principal attention and it became a concern of social geographers. Tran and Roosvelt (1986) came with an interesting attempt to identify the causal predictors of subjective wellbeing among the Vietnamese refugees in United States. Path Analysis is used to test the causal influence of variable posited to affect wellbeing in the hypothesised causal model. Social support, Social interaction anxiety, family income and marital status were found to have significant direct effect on refugees' sense of wellbeing.

The quality and vulnerability of physical environment are also very crucial in the social wellbeing. Davis (1986) measured quality of life giving focus on the small island developing nations on the Indian Ocean. The physical quality of life index is used to evaluate wellbeing. It employs Infant mortality, life expectancy at age one and literacy which are indicative of a wider range of social and physical attributes. The result of Indian Ocean islands group is seen with Reunion, Mauritius and Seychelles consistently performing very well while Maldives and Comoros perform poorly

Subjective quality of life is also very important in urban geography that how the people perceive the environment and quality of life and what is the wellbeing in their opinion. Desai (1986) has given an account of people's perception of urban environment in terms their socio economic background and experience. She expressed the issue of environmental perception both from qualitative and quantitative points of view by using several indices of residents' perception in different areas of Ahmadabad city. The author also observes that the level of perception of environment is different among various communities according to their awareness and knowledge about environment.

Meier (1986) in his work denoted that the world cities as the pace maker among urban settlement. He attempted to put forwarded a new model to explain the quality of urban life using the examples of world top cities. The author tried to crystallize the facets of quality of life under income, prestige, services culture and glamour. Twenty-two scenarios were constructed which spelled out possible changes in the societal relationship and economic consideration which could directly affect the city environment.

Overall assessment of urban environment and the social wellbeing outline the liveability in cities in micro level administrative boundary. Some area will be liveable and other may be less comfortable to live due to either physical and social environment or both. Pacione, (1990) tried to conceptualize the urban liveability that is urban environmental quality. Author simplified the complex concepts through the wide literature references and the appropriate explanation and addition. In order to draw out the quality of urban environment author emphasized the urban crowding, urban legibility, site design and social behaviour and the urban sub areas. The study included the objective and subjective measure to obtain a full appreciation of environmental quality that considered city on ground and city on mind.

Fakhruddin (1991) initiated the city level studies of the quality of urban life in India with the usage of sound methodologies and techniques like factor analysis. The author attempted to outline the residential structure and the quality of urban life taking the case of Lukhnow city. The author organized the entire study on five important factors –material and housing condition, territorial stress, amenities and infrastructure, health and survival and lastly education and recreation. It explains the residential characteristics in relation to the environmental quality with the emphasis that the evolution of residential pattern and structure in the city are the outcome of the persistence of traditional structure, colonial modification and the economic prosperity.

Pothna et al. (1992) tried to outline the socio economic condition and the access to the public amenities among the slum dwellers of Vishakapattanam city. By the intensive sample survey in the slum dwellers and also by secondary data they drawn out the distribution of slum population in terms of standard of amenities and compared the same with the non-slum population to show the relative slandered of amenities. They suggest that the market intervention to regulate the flow of services to the poor is essential and the local government should care to mobilize the resources necessary to improve the quality of life of poor slum dwellers from within the system.

The inequality in the resource allocation, ethical and social injustice and allied differences in social wellbeing became the focus of the urban geography in 1990s. Smith (1994) conceptualized the social justice in terms of multiple elements and set of theories. The studies is based on first hand information and it covers racial injustice in United States, inequality under socialism in east Europe and the prospects for social justice in post apartheid South Africa. By portrayal of those people who have no secure place or defined territory, he draws the attention to the justice of market led society, the ideas of egalitarianism and the principles of justice. And the study emphasis that geographers can make to the understanding of social justice in a complex and rapidly changing world.

Felce and Perry (1995) proposed a model of quality of life that integrates objective and subjective indicators, a broad range of life domains, and individual values. It takes account of concerns that, externally derived norms should not be applied without reference to individual differences. The work also allows for objective comparisons to be made between the situations of particular groups and what is normative. Considerable agreement exists that quality of life is multidimensional. Coverage may be categorised within five dimensions: physical wellbeing, material wellbeing, social wellbeing, emotional wellbeing, and development and activity.

Up to 1990s several models and methods to assess the quality of life and wellbeing are emerged among the researchers, academicians and planners. The appropriate use of the specific methods became a matter of discourses in this time. Raphael et.al (1996) reviewed six approaches that consider quality of life and health. They are (a) health-related quality of life; (b) quality of life as social diagnosis in health promotion; (c) quality of life among persons with developmental disabilities; (d) quality of life as social indicators; (e) the Centre for Health Promotion (University of Toronto) model, and (f) Lindstrom's quality of life model. Each approach is considered as to its emphasis on objective or subjective indicators, individual or system-level measurement, value-laden or value-neutral assumptions, and potential relationship to social policy and social change goals. The links among the social indicators, quality of life, and health promotions areas are also examined.

The effect of political system and independence is also assessed in the area of social wellbeing that this strategy is became a frame of development and prosperity. In the study of Wickrama and Charles (1996) the level of political democracy is hypothesized to have an independent positive effect on social wellbeing irrespective of either level of economic development or level of disarticulation of economies of developing countries. In addition, political democracy is hypothesized to buffer the negative effects of disarticulation on social wellbeing. The study proposes two possible effects of democracy on social wellbeing. First, there is a potential independent direct effect of democracy on social wellbeing over time. Second, there is a potential moderating effect of democracy on the relationships between dependency conditions and social wellbeing.

A further step in the subjective measure of wellbeing came with the incorporation of the definitions and opinions of the residents about the quality of life and wellbeing. Zenher (1997) attempted to measure objectively the subjective concept of the quality of life in new communities. Different strategies are used to analyze the resident's over all life satisfaction. A unique aspect of this study is the incorporation of residents' own definitions and perceptions of the factors that influence the quality of life. He used the primary data with nineteen community attributes under housing, community facilities, physical environment, social environment and work transportation and living cost to drown out the situation of thirteen different communities and classified them to different categories.

Household based indoor environment is also very important in quality of life and urban environment as it directly affect the health and psychological conditions of residents. A study of urban oriented household environmental problem is conducted by Atiqur Rahman (1998). The study is designed to provide a coherent assessment of household environmental problems faced by different income groups. By sighting the case of Aligarh city the author putts a detailed account of different aspects of the household environment and related health problems. The summery of the study is that there is positive correlation between the household environment and health condition of the people lives in.

A study of urban unemployment in Kerala is conducted by Prakash (1999) taking Cochin City as case study by using primary data collected through the technique of stratified random sampling, realized that, the rate of urban unemployment in Kerala is very high with the national average. The study also shows that the incidence of unemployment among the educated young is due to the excess supply of educated young labour force in one hand and the small size and slow growth of organized sector leading to low demand for regular employment on other.

Similar to social inequalities the regional imbalances in the level of living in cities are also be researched and studies. Wei and Fan (2000) came up with an interesting study of regional inequality in China. They considered the important factors led to the regional imbalance as the local agents and the foreign investment. They depicted the inequality in terms of inter regional and inter country level. The main arguments can be noted as the factors and agents of regional growth must be analyzed in the light of the complexities of the economy, respective historical, geographical and political contexts.

Personal security and free from the fear of crime is integral to psychological wellbeing. Evans and Fletcher (2000) came with a study of fear of crime in English midlands. It examines the spatial and social distribution of the fear of crime and the relationship of such fear with different aspects of the environment. The authors used 58 variables to analyse the situation. It considers both causes of the fear of crime and the associations shaping vulnerability. They associated the fear variable with age, job status, marital status, and gender. But the main defect of the study is that they ignored the expression of crime fear in terms of urban and rural lands which have paramount significance in such studies.

Urbanization also led to the deterioration of the physical environment and ecosystems present in and around the city. Jochem and Waals (2000) committed to outline the negative effect of the urbanization on environment. Four categories of environmental effects discussed, they are emission of carbon dioxide and oxides of nitrogen caused by automobiles, energy uses in houses, noise odour, and local air pollution and fragmentation of natural areas. The study is derived two approaches that, measurement of the behaviour factors and environmental pressure, and second one is the scenario studies. They view the relationship between compact urbanization and environmental quality as full of complexities.

Valuation of quality of human life, measurement of social wellbeing and evaluation of policies for sustainable development is conceptualized by Dasgupta (2001) in the book Human Wellbeing and the Natural Environment. The author focussed on policy reforms and developed a measure of social wellbeing in time and coined a population theory. It has a comprehensive coverage in concepts and issues of human wellbeing and accounting for nature and sustainable development. The author broadened the source of human wellbeing beyond the human and manmade capital and knowledge base to include, the most importantly, the natural capital or natural environment. It is resource allocation mechanism which precisely determines the inter-temporal flow of social well being and the associated dynamic paths of the concerned stocks which describe the composition of wealth of a society.

Recent studies reveal that the religion and tradition is a attribute in the moulding of the urban social environment which leads the wellbeing of the society and locality. Concept of human wellbeing that integrates economic and non-economic aspects of life is elaborated by Tomer (2002). According to the author, philosophers, humanistic psychologists, and religious traditions have been very helpful in pointing out the true non-economic potential of human life. To adequately assess a person's or a society's wellbeing, it is necessary to consider both people's ordinary (or lower) functionings and their overall (or higher) functionings. To raise the societal wellbeing it requires capital formation, particularly investment in personal and social capital.

There must be a common space or bridge between the environmental quality and social wellbeing and the policy makers must tackle this problem. Based on experience in the environmental field, Brown (2003) suggests that one of the major challenges is to bridge the divide between the environmental quality/wellbeing/quality of life specialists and the planners who make urban policy and who shape our physical and social environments. In order to bridge the divide between environmental quality specialists and development practitioners following strategies have to be adopted. (a) Understand how the development process functions; (b) understand the language, the tools and the thought processes that are used by the players in the development process; (c) assist the development manager to develop scenarios; (d) and do not impede, too much, the processes that the development manager currently uses.

On the commencement of the 21st century social geographic studies are revolve around the concept of man and environment relationship, and the working and the extension of the mutual interaction and relation. Pacione (2003) again approaches the issues of urban environmental quality and human wellbeing from a social geographic perspective. Seeking to understand the nature of the relationship between people and their environments is the perfect geographical question that, in the context of the built environment, may be interpreted as the degree of harmony or dissonance between city dwellers and their urban surroundings. The author addresses the major theoretical and methodological issues confronting quality of life research, and presents a five dimensional model to guide work in the field. Two case studies presented by the author demonstrate the value of a social geographic perspective on quality of life. Finally, several outputs of quality of life research of relevance for social scientists and policy makers seeking to improve living environments in the contemporary city are identified.

Helliwell and Putnam (2004) explored the social context of subjective evaluations of wellbeing, of happiness, and of health by using large samples of data from the World Values Survey, the US Benchmark Survey and a comparable Canadian survey. Social capital, as measured by the strength of family, neighbour- hood, religious and community ties, is found to support both physical health and subjective wellbeing. The authors confirm that social capital is strongly linked to subjective wellbeing through many independent channels and in several different forms. Marriage and family, ties to friends and neighbours, workplace ties, civic engagement, trustworthiness and trust: all appear independently and robustly related to happiness and life satisfaction, both directly and through their impact on health.

The poverty is an urban sociological assignment which is yet to be embark upon even in the cities of developed nations. Vaughan et. al. (2005) have taken interest to study the special causes of poverty and the persistence of it on London city. They drown out the poor areas in relationship between spatial segregation and poverty; it shows the space can itself be considered as a factor in the geography of poverty and how the morphology of the streets can have an impact on people live in.

Stewart (2005) had an attempt to outline the dimensions of wellbeing across European Union regions. Using data from a range of sources, the paper examines the association between unemployment, gross domestic product (GDP) and a number of alternative wellbeing indicators in five domains - material welfare, education, health, productive activity and social participation. It finds that GDP per capita is not a good proxy for wider regional wellbeing within a country, but that the regional unemployment rate performs reasonably well.

The globalization, neo-liberal urbanization and biased political agendas are again changed the life style and reinforced the problem of people. Brand and Thomas (2005) make an important contribution to understand urban environmentalism as an ideological form. The authors enquire why the cities have embraced environmental issues with enthusiasm. The work locates urban environmentalism within current debates on globalization and neo-liberal urbanization, and critically outlines the political success of urban environmental agendas in the postmodern condition of risk and individualization. These themes are subjected to theoretical critique and methodological exploration through Marxist analysis, discourse theory and a relational understanding of urban environmentalism. The approaches then applied through in-depth second city studies in contrasting development contexts: Birmingham, Lodz and Medellin.

Royuela and Artis (2006) analyzed the convergence of quality of life in Barcelona province. It is based on the concept that, the individual wellbeing depends upon several quality of life factors and that convergence across the territory is as important at the local level as at regional or national level. The usage of the vast indices and variables are ideal, they used three main components of quality of life that is individual opportunities for progress, index of social equilibrium and community conditions of life. Under these components they used 18 sub indices and 63 variables in different angles of well being.

Srinaivsan (2006) came with the interesting study of the equity, accountability and environmental concern in the solid waste management in Chennai city. The paper explores the issues related to effectiveness, role of the urban local body, private agencies and a civil society organization engaged in solid waste management and the relevance of effective policy frame work. The entire study is based on qualitative research methodology and involves field observation and interaction with residents and officials of all three agencies that is public, private and voluntary. Such things are very important that if they are not treated properly it will deteriorate the physical environment of the city which can lead to the health problems and hence the wellbeing of the residence.

Andreas and Ortega, (2006) came forward with the extensive study on the Spanish region in terms of quality of life and economic convergence. The authors examined the quality of life by the alternative composite indicators in the context of UNDP's human development indices. Three per capita income measures and seven indicators of quality of life were studied. The author's major findings may be summarized as there is a high correlation between levels of income and achievements of human development and regions with similar gross value added per capita can have very different achievements in HDI

Seemin and Firoz Khan (2007) studied the quality of urban environment taking the case of Saharanpur city. The authors depicted the dimensions of quality of environment in the study area by six principal components of different dimensions i.e. general environmental status, territorial stress, water supply and hosing facilities, waste production and the open and green area. The study is based on primary data collected by sample survey. The main findings of the study is that the liveability of urbanites is based on economic status, those at bottom on hierarchy bears the burden of under development that is the inner and middle zones of Saharanpur city enjoys relatively better living condition and outer city is worse in some parts.

In most of the cities all the facilities are available but the access of the whole inhabitants to the aforesaid facilities is a problem due to several reasons like economic and social status. Valhov et al. (2007) joined the urbanization development and slum as the determinant of urban health. Four broad categories of living condition i.e. population composition, physical environment, social environment, and availability and access to health and social services- have put as the principal determinant of urban health. The social determinants that frame the urban setting given stress especially the density and diversity of population, social and human resources and political structure.

The health factors open to get a paramount importance in wellbeing in recent years. The social environment is a key to understanding the way in which cities affect the health of populations. Ompard et al. (2007) briefly summarizes a few key social determinants of health and their measurement, and also considers methodological tools and some methodological challenges. The concepts presented in the paper are broadly applicable to a variety of settings: developed and developing countries, slum areas, inner cities, middle income neighbourhoods, and even higher income neighbourhoods. However, the focus is given on some of the more vulnerable urban populations who are most profoundly affected by Social determinants of health.

Place of residence, gender, race/ ethnicity, education, and socioeconomic status are taken as key determinants of social determinants of health.

Bleys (2007) appeared a different approach to measure welfare, that is, the existing national accounts, the access of people to certain basic need and the mental stages of subjective wellbeing. The index of sustainable economic welfare is given emphasis in this study. The author tried to overview that for every society there seems to be a period in which economic growth bring the quality of life only up to a threshold point, beyond which it may be deteriorated . A case study presented for India by available data on different types of alternative welfare measures.

Now the discourses came together to gauge the quality of researches, quality of indicators and the methodologies involved in the social wellbeing researches and the utility and applicability of the same. Moller (2008) outlined the quality of life research in developing countries sighting the case of South Africa. The research methods and the selection of hypothesis and indicators have been used in an attracting way. A set of indicators for subjective quality of life and objective wellbeing is used and correlated them. Socially excluded population and illiterate population given stress and also included them in the interview to get actual picture of wellbeing. The result of the work indicates that there is a vast gap between the level of satisfaction of blacks and whites, and rich and poor in society. The influence of optimism is not avoiding in the negative correlation with current happiness and satisfaction.

The wellbeing and quality of life acquired global dimension and common criterion and concepts are emerging like the human development index. The individual and social wellbeing within the broader context of global human problems and planetary wellbeing is conceptualized by Sandra Carlisle et.al (2009). The escalating growth of materialistic and individualistic values of western societies is associated with a growing sense of individual alienation, social fragmentation and civic disengagement and with the decline of more spiritual, moral and ethical aspects of life. Taking together, these multiple discourses suggest that wellbeing can be understood as a collateral casualty of the economic, social and cultural changes associated with late modernity. However, increasing concerns for the environment have the potential to counter some of these trends, and in so doing could also contribute to our wellbeing as individuals and as social beings in a finite world.

Over all, if the whole condition of a society is well, there may be inauspicious ill-being caused by the inborn or hereditary disabilities for an unavoidable population, they also have in thirst of care and satisfaction. Jin-Ding Lin et.al (2010) in their study are to explored the wellbeing perception and its determinants of caregivers who caring for people with disability. The authors employed a cross-sectional, self-administrative structured questionnaire survey. Wellbeing was measured using two scales which included Subjective Happiness Scale (SHS), and Satisfaction with Life Scale (SWLS). With respect to the determinants of respondent's SHS in a multiple linear regression, the authors found the factors of perceived health status and SWLS were variables that can significantly predict the SHS. The study suggests the health and welfare service authorities should pay attention to the wellbeing profile and determinants of caregivers who working for people with disability to improve their quality of life.

Thus the study of urban environment and human wellbeing is obligatory at different levels as an emerging problem associated with new urban setting with a highly fluctuating urban way of life having consumer oriented actions and mentality. It is very clear that the deterioration of urban environment is other face of the urbanization. New analyst tools for identifying the actual threats and working plans for the minimization of the effect of urbanization on the environment must be evolved through quality researches. So there must be a more sophisticated multidisciplinary coordination to distinguish the different aspects of economic, ecological, psychological, geographical and medical aspects of the urban places and urban dwellers.

3

CONCEPTS OF URBAN ENVIRONMENT AND SOCIAL WELLBEING

Geography can be variously defined as the science deals with the spatial organization, locational analysis, man environment relationships and areal differentiation. Geographers are professionally interested in how and why one piece of land identical from another with respect to both its physical environment and the activities of man (Smith 1973, p. 10). The city is not only the physical environment but there is another environment that some times more important than the physical conditions for the residents, that is social environment. So the analysts and planners must pay attention to the city social environment along with the physical environment.

Physical and Economic planning of cities has been an established part of the governmental scene since 1930s. But Social planning has been openly recognized only more recently, and then it has proceeded under a cover of confusion which has prevented public debate on its scope and intention (Smith 1973, p. 56). The industrial revolution led to the people organizes in abundance in specific areas and it commenced the urbanization throughout the world and the process is on the work.

It is the first time in the history of Humankind a majority of the world's population lives in urban places and the number of urban dwellers and level of global urbanization are growing at an accelerated rate. The quality of the urban environment as a living space for the inhabitants, therefore, is an issue of fundamental concern for academic researchers, policy makers and citizens themselves (Pacione 2003a, p. 1)

Socio-spatial variations in urban environmental quality and human wellbeing are established characteristics of the city life. The problems of living in the contemporary city drew the national and international research attention to these socio-spatial divisions in urban environmental quality and human wellbeing (Pacione 2003a, p. 1). Even in the recent days due to manifold reasons the city and urban society on which the pulse of world is resting, persists as an enigma that to be tackled. Most of the sociological and technological transformations fire up from the urban areas and make ripples throughout the world's society.

Cities are of greater complexity today than ever because of its size and population density. There is a more important reason for this urban complexity that is urbanism now dominates civilizations, and cities have assumed a commanding position in whole world. So the study of the city perforce has become the study of contemporary society. It is the urban societies that control the world's destiny as the centres of decisions and the triggers for social change (Reissman 1964, p. 3)

As an important section of the modern society at whom rest of the world depends for the technology and even for ideology the urban population deserves the care of policy makers to tackle the internal problems of the cities especially in social context. 'The development of humanistic welfare geography is catalysed the demands of public policy. Improving the wellbeing of specific neighbourhoods, cities, or regions requires some means of accurately describing spatial variation in wellbeing so that policies can be formulated and later assessed efficiently' (Knox 1975, p. 4)

The city is a relatively large, dense and permanent settlement of socially heterogeneous individuals (Nottridge 1972, p. 37). Heterogeneity is the life blood for the existence of mankind so that the symbiotic relation could be kept to fulfil the wants and needs of all sectors in a society. The heterogeneity leads to the conflicts if not managed it in a proper way by both the individuals of community and the political authority.

3.1 Urban Environment

Urban environment understood as those entire external phenomenon prevail in the urban area that facilitate or malignant to the community or society present there. Generally the most of the people understand the word environment as only ecological or natural environment, but urban environment include ecological as well as social environment. 'Environment is the aggregate of external conditions that influence the life of an individual or group, specifically the life of man; environment ultimately determines the quality and survival of human life' (Detwyler et al. 1972)

Cities are open and dynamic ecosystems, which consume, transform and release materials and energy. They develop, adapt, and interact with other ecosystems. The quality of life in cities is a function of interactions with its different components such as social equity, income, housing, healthy environment, social relations and education (EEA 2009). The city must be treated as only one of many social institutions such as kinship, religion, and subsistence activity for cross-cultural comparisons and analyses of urban environment (Fox 1977, p. 17)

Urban geography greatly concerned with form in 1960s and 70s, and then the classical relationship between urban land use and morphology with emphasis on site, location, and physical conditions, is receded into the focus. Finally geographical investigations began to consider the internal features of the structure of an urban system, and these features which are powerful among the environmental forces conditioning urban life, belong to the 'built environment' of the place and the socioeconomic characteristics of the population (Harrison and Gibson 1976, p.14). But the scientific world not fully satisfied with this type of studies. The germination of humanistic and radical thoughts in geographical explanations brought the habitat and wellbeing factors in to the focus.

The new innovations in the geographical investigations such as Quality of urban environment came after 1970 when concern with human habitat became subject of more focussed research within the older traditions of 'quality of life', 'social indicators', 'human development' etc. as a dissatisfaction from gross economic measures of production and consumption as true indicants of human wellbeing and it developed a new approach to the human wellbeing (Harrison and Gibson 1976, p. 19). These new indicants are more concerned with the residential environment and the land use system within the city space.

Residential environment is the outcome of several factors that not only related with natural environment but also with the social environment. The special interest on environmental quality has emerged as a key area for research in urban social geography, especially for research undertaken from an applied or problem oriented perspective. Accordingly, within urban social geography considerable efforts have been directed to assessing the quality of different residential environments (Pacione 2003b, pp. 19-20). Urban land use is also very important; the urban area with lot of industrial area will have a deteriorated physical environment and the concentration of residential land use may have an influence in social environment. There is a fairly direct relationship between the 'natural' environment and land use. 'At the stage of urban expansion, residential sectors were moving into human landscape where the social values had replaced those of soil and site rather than that of physical landscapes' (Harrison and Gibson 1976, p.14)

Moreover, it is very common that the direction of urban development and land use expansion have largely controlled by the physical presence of natural barriers like river, sea and mountain, and the pattern of settlement followed it. The study of the urban environment was evolving in rigid fashion along a line that increasingly subordinates the physical features of the milieu, either natural or manmade, lagging the social and cultural ends behind the economic and technological means (Harrison and Gibson 1976, p.16)

There is a paradigm shift in understanding of environment especially urban environment. Several scholarly discourses and researches have covered this issue. In the recent years, the environment is a totality of quality of individual's and society's life along with the physical environment. After the huge social violent and social dilemmas of the previous decades the environment was seen as a mean to re-establish a sense of quality of urban life at community and individual levels, expressing both the practical concern and the intellectual optimism (Brand and Thomas 2005, p.1)

The geographer who perceives the urban community as bounded in certain area, or locality, or in an urban place uses a spatial approach to study the urban phenomenon. The ultimate reasons of this study is to understand the organization of the place in order to serve ultimately the general wellbeing of the society by considering the urban spatial aspect (Harrison and Gibson 1976, p. 19) According to Human ecologists the city is primarily a natural environment. Within it they expected to study the effect of ecological unit in which pattern and process could be notice by the same technique and from the same perspective as those in nature (Reissman 1964, p. 99)

Ecological environment is one of the main factors which have an effect on the way of life and economic activity in both rural and urban area. Nature of economic activity is controlled by natural resources, physical conditions and the relative location. As Sengupta's opinion Human being depends on Ecological environment for their survival and development which includes an array of natural assets of minerals, sunlight, fossil fuels, soils, water, forests, water sheds, biodiversity, oceans, atmosphere etc. which provide support to human living including functioning of economic system'(Sengupta 2002, p. 4289)

Urban Community environment is not just a part of the urban environment, but it is also a concrete reflection of social, ethnic, cultural and material progress (Liu 2000). Urban man have several needs that to be fulfilled for their happy and well living. It may be mental or psychological as well as physical. Some needs are essential for their personnel happiness or quality of life and some other for the better living for their whole urban community. Two broad categories of human needs may be noted, biological needs essential to survival of the urban population and cultural requirements necessary for city functioning and growth.

The biological needs of man in the city are essentially the same as he requires in any situation, such as food, energy resources, space, water and shelter materials which generally draws from an extensive hinterland due to the shortage by high population densities. Space might be the biological requirement which obtained little attention and the urban ecosystem most seriously failed to provide. The culture is the totality of man's ways of living developed by human groups and transmitted from one generation or group to another which include political organisation, economic system, technology, transportation and communication, education and information, social and intellectual activities etc. It is very essential to the urban as a whole for their better development and competency (Detwyler et al. 1972, pp.14-17)

There is a divide between environmental quality specialists and development practitioners. In order to bring them together the environmental quality specialist must understand the functioning, language, tools and the conceptual elaborations of the development process. The former must assist the later to develop scenarios, particularly novel paths of action that could shift the development activity in preferred directions. The environmental quality input must not impede the processes that the development player currently uses (Brown 2003, pp. 87-88). This divide must be reduced to acquire a mutual understanding to become the planning effective and environmental friendly.

Environmental quality considered all the aspects of overall life satisfaction such as quality of housing, interaction with neighbours, public transportation, and health services (Raphael et al. 1996, p. 78). Overcrowding, inadequate housing, inadequate access to clean water and sanitation, growing amounts of uncollected waste, and deteriorating air quality are serious problems in cities and there must be effective and timely action to tackle these. These phenomena will seriously affect and deteriorate the urban liveability and quality of life with varying multitudes. One of the most obvious urban features, after size and congestion, is variety. The city is a place of contrasts, an environment of extremes. A good relation between urbanites and urban communities must be formed to get an improved social environment. Cities are communities only because they command allegiance, social consensus, and belief, and even, at times, a civic spirit (Reissman 1964, pp. 4-13)

Physically and socially cities are still an imperfect entity. It has never been a perfect environment. Yet we should not let our sensitive awareness of urban problems totally darkened. It seems likely that such imperfection is an inevitable feature of urban complexity. Throughout the history of urban society there have been recurrent crises, and probably never more than in the last hundred years, as we have realized that something could be done about those problems, that they were neither natural nor inevitable. (Reissman 1964, p.10-12)

3.1.1 Urbanization and Social Change

Urbanization associate with social and technological development and a better society to transmit the education and technology through it. Urbanization is Social Change on vast scale. In recent years the shift from an agrarian to an industrial society has altered most of the aspect of social life. The family reduced the limit of its allegiance and modified its relationships. The economy was drastically altered in style, purpose and demand. Education was remodelled to fit urban and industrial needs. Politics occupied a different arena than before, with new participants, rules and objectives. The motivation of urbanization upon society is such that society gives way to urban institution, urban values, and urban demands (Reissman 1964, p.154)

Urbanization in the perspective of social change means not only the transformation of rural, agricultural, or folk society, but also the continuous change within the industrial city itself. Urbanization does not stop but continuous to change the city into ever different form (Reissman 1964, p.156). People in cities have restricted contact with others than those lived in a less crowded environment. To some extend loneliness is a problem in cities. Furthermore, the city increases the ease with which one can restrict our human contacts to people of his own kind. In cities very specialized subcultures develop whose sympathy and experience can be much more limited than those of the country dwellers (Harrison and Gibson 1976, p. 293).

In the same way the internal structure of towns and cities is modified by changing characteristics of size form and function. Generally the socioeconomic changes develop faster than the physical and built-up environment. The renewal of the environment attempts to catch up, but as a whole lags behind the needs; change has occurred quickly in last century. Size, form, and function concur to extend rapidly the space occupied by urban agglomeration (Harrison and Gibson 1976, p.15). 'The quality of the environment is a key indicator of a successful city centre. For its part seduction was aimed at the city's internal population. A revitalized and attractive city centre environment would not only edify a new collective image but also, it was argued, offer new opportunities for cultural consumption and urban social life' (Brand and Thomas 2005, p.117-118)

Inter-personal relationships benefits people both emotionally and physically. The quality of a neighbourhood, including good quality public space, is important for motivating contacts and strong social ties among neighbours, which might be of greater importance for low-income people (Putnam 2000). There are considerable links between urbanization, the growth of cities, and increasing crime rate. Crime and delinquency have become regarded as urban phenomena because of the regular annual increase in crime rates, especially those for violent offences that became a feature of some societies. Crime remains one of the least understood social problems, particularly in the key context of causality and prevention (Herbert and Smith 1979, p.117). It is clear that urbanization leads to social changes at different scales in association with many other factors like economic development, educational awareness etc.

3.2 Social Wellbeing

The study of social wellbeing is an outcome of deteriorating social environment due to the overcrowding and less social contact which came in existence because of the enhanced agglomeration of residential and industrial establishment or simply urbanization. Social wellbeing is associated with the total condition of individual and community life. The satisfaction with ones social or community environment is very important for his mental and physical development. 'Wellbeing is a multidimensional concept that comprehends many criteria quality of life and many life domains'. (Andrews 1983)

Several definitions are came forth to conceptualize the social wellbeing, quality of life, level of living which are more or less resembles in nature but slightly differs in dimension. According to Paul Knox wellbeing is the satisfaction of the needs and wants of the population, and the needs associated with different elements of wellbeing may be resolve in different ways. The consumer needs may be determined through conventional supply and demand analysis, recreational needs through relative deprivation analysis, housing needs through statistical analysis and medical care needs through resolutions of expert's opinion (Knox 1975, p. 6-7)

A well society is one in which people have adequate income for their basic needs of food, clothing, shelter, and a reasonable standard of living; people will not live in scarcity. For any individual his status and dignity will be respected, and will be socially and economically mobile. Decent education and health services will be accessible to all, and their use will be reflected in a high level of physical and mental health. People will live in suitable houses in decent neighbourhoods, and will enjoy a good quality of physical environment. They will have reasonable leisure time and access to recreational facilities, including culture and the arts. The populace will show a low degree of disorganization, with minimum personal social pathologies, little deviant behaviour, low crime incidents and high public order and safety. The family will be a stable institution. Individual will be able to participate in social, economic, and political life without any discrimination based on the race, religion, ethnic origin or any other cause (Smith 1973, p. 69).

Quality of life can be termed as the sum of a variety of objectively measurable life conditions experienced by an individual. These may incorporate physical health, personal circumstances, social relationships, functional activities, and wider societal and economic influences. Subjective reply to such conditions is the domain of personal satisfaction with life (Felce and Perry 1995, p. 54). Raphael defines quality of life as: 'The degree to which a person enjoys the important possibilities of his/her life. Enjoyment covers two meanings: experience of subjective satisfaction and the possession or achievement of some characteristic or state. Possibilities reflect the opportunities and limitations each person has' (Raphael 1996, p. 80)

Those who concerned with wellbeing must be satisfied with comparative rather than absolute measures. Until we standardize collective preferences and priorities, such measures must be based on those categories of needs and wants that are seen by academics and administrators to constitute wellbeing (Knox 1975, p.7). Generally there are three types of wellbeing which are mutually related. They are Physical wellbeing, Social wellbeing and Emotional or Psychological wellbeing.

Physical wellbeing or material wellbeing subsumes the health, fitness, physical safety, finance or income, quality of the living environment, and privacy, possessions, meals or food, transport, neighbourhood, security, and stability or tenure. Social wellbeing includes the quality and breadth of interpersonal relationship with the family and relatives in the surrounding people and friends. Community activities and the level of community acceptance or support together reflect a similarly strong concern for community involvement. Development and activity is concerned with the possession and use of skills in relation to both self determination and the pursuit of functional activities (work, leisure, education) and productivity or contribution. Emotional wellbeing includes affect or mood, satisfaction, or fulfilment, self-esteem, status/respect, and religious freedom and faith. (Felce and Perry 1995, p. 60)

The social wellbeing is largely depends on the level of material possessed or accessed as that supports all other parts of life and maintain social status. The quality of people of a society depends on the resource allocation mechanism which is characterised or determined by the institutions, motivation and culture of peoples of the society. Resource allocation mechanism is the management of assets of all kinds of capital stocks. It determines the inter-temporal flow of social wellbeing and the associated dynamic paths of the concerned stocks which describe the composition of wealth of a society (Sengupta 2002, p. 4289)

As many other phenomena the wellbeing can be studied or explained in several ways. It is differing from scholars to scholars and discipline to discipline. According to Chan 'There are two distinct orientations to the study of wellbeing- the 'grass roots' approach and the 'ascriptive' model. The approach bank on people's judgments of needs and satisfactions according to individual life experiences and expectations is known as grass roots approach. Researchers analyses the collected public opinions, and construct a subjective framework on quality of life that focuses on individual subjective feelings like degrees of satisfaction. (Chan et.al 2005, pp.261)

There are several factors which influence or regulate the social wellbeing and the well-off of the society. The most important factor among them can be termed as the population characteristics. There is a strong and direct connection between levels of wellbeing and characteristics of populations. Changes in several population parameters are influence the levels of wellbeing, similarly, changes in levels of wellbeing affect various aspects of population. Even though demographers have developed an advanced discipline around concepts involving rates of fertility, mortality, migration, marriage, divorce, dependency, and the like, they given relatively little attention to link the data they developed to communicate about wellbeing. Social epidemiologists have explored how some aspect of wellbeing like emotional and physical health correlate to certain population trends, but the range of wellbeing phenomena examined has been very limited. (Andrews 1983)

Being, Belonging, and Becoming are the three life domains. Being indicates "who one is" and has three sub-domains: Physical Being incorporates physical health, personal hygiene, nutrition, exercise, grooming, clothing, and general physical appearance. Psychological being covers the person's psychological health and adjustment, cognitions, feelings, and evaluations concerning the self such as self esteem, self-concept and self-control. Spiritual being contains the personal values, personal standards of conduct, and spiritual beliefs which one holds.

The Belonging domain be connected with the person's fit with his/her environments and also has three sub-domains. Physical Belonging involves the person's connections with his/her physical environments of home, workplace, neighbourhood and community. Social Belonging consist of links with social environments and involves acceptance by intimate others, family, friends, co-workers, and neighbourhood and community. Community belonging includes access to resources such as adequate income, health and social services, employment, educational and recreational programs, and community events and activities.

Becoming denotes the purposeful activities carried out to express oneself and to achieve personal goals, hopes, and aspirations. Practical Becoming describes day-to-day activities such as domestic activities, paid work, school or volunteer activities, and seeing to health or social needs. Leisure Becoming describes activities that promote relaxation and stress reduction. Growth Becoming activities promote the maintenance or improvement of knowledge and skills and adapting to modification (Raphael et al. 1996, pp.80-81).

Quality of life can be perceived as the quality of life conditions, satisfaction with life, and the combination of both, and it is an intangible concept approachable at varying levels of generality from the assessment of societal or community wellbeing to the specific evaluation of the situations of individuals or groups (Felce and Perry 1995, p. 51). Lindstrom's quality of life model considers four spheres- Personal sphere, interpersonal sphere, external sphere and the global sphere. The Personal sphere covers physical, mental, and spiritual resources. Interpersonal sphere includes family structure and function, intimate friends, and extended social net works. The External sphere includes aspects of work, income, and housing. The Global sphere includes the societal macro environment, specific cultural aspects, and human rights and social welfare policies. (Lindstrom and Spencer 1994)

Level of living is a fundamental concept for the assessment of the development and the development planning. The purpose of development is to improve the conditions in which people live, and the level of living is supposed to be a quantitative expression of these conditions. Only material wealth cannot bring wellbeing to the individual or society. But it is a totality of several factors of material, social, political, psychological, biological etc. 'Quality of life is not necessarily a simple function of material wealth, but it includes other factors like social, political and environmental health of a nation. This realization has led to the search for indicators, other than those based on GNP, that will reflect more adequately the overall health of a nation and the wellbeing of its citizens'. (Pacione 2003b, p.19)

The living condition as mentioned above is not only the physical condition but include all the life domains. Knox explaining the living condition and social wellbeing much vastly and simply to understand the domains of life which influence the wellbeing. 'The level of living of person or resident within a given geographical area is constituted by the over-all composition of housing, health, education, social status, employment, affluence, leisure, social security, and social stability aggregately exhibited in that area, together with those aspects of demographic structure, general physical environment, and democratic participation' (Knox 1975, p. 31)

3.2.1 Religion and Wellbeing

Religion has been an important aspect of human activities and a way of life in most of the communities both rural and urban region. Religion is related with life and life's various dimensions. It has an influence on the entire life of man as a person and a part of family and community. 'Religion provides an important basis for ideas about wellbeing specifying through teaching and practice what it mean to live well, as an individual and as a community. It is also widely understood as a source of wellbeing for its adherents, providing comfort in times of trouble, offering a frame work of meaning to make sense of life's variations, and providing a community that gives social support and confers identity through a sense of belonging' (White et al. 2010, p.1)

Religions and spiritual traditions inevitably incorporate ideas concerning what the good life is, how one should live one's life, and the rewards (and penalties) one can expect in this life or possible future lives. Almost all religions specify some ultimate or ideal state that their members aim for such as enlightenment or oneness with God. In a sense, this state provides the member with the ultimate of wellbeing, but since typically only a very small proportion of the religion's members achieve the ideal, the most important thrust of a religion is its guidance on how to live one's life, and thus, how to achieve a measure of increased wellbeing on the path (Tomer 2002, p. 34).

In the case of Hinduism, there are four basic aims in life that contribute to wellbeing. The ultimate aim of Hindus is self-liberation or self-realization (*moksha*). The other three are, enjoyment of the fullness of life or pleasure (*kama*), means of life, success, prosperity or wealth (*artha*), and virtuous living, righteousness, guide to action, or moral and ethical duty (*dharma*) (Koller 1985, pp. 42-46). Hindus can make significant progress along this path, and increase their wellbeing, by bringing *artha, kama*, and *dharma* into the right relationship.

With respect to wellbeing, the essence of Buddhism is quite similar to Hinduism. According to Buddhism, there are four factors contributing to happiness, wealth, worldly satisfaction, spirituality, and enlightenment. In regard to spirituality, "your state of mind is the key". It is important to have the "right mental attitude" and maintain a "calm and peaceful state of mind". Buddhists, similar to Hindus, seek to reach the ultimate state of enlightenment, through disciplined practices that remove the "dust from the lamp", alleviate suffering, and allow the Buddha nature to shine through (Lama and Cutler 1998, pp. 24-25)

It is very easy to characterize the subject of wellbeing in Christianity. Christianity is distinguished by a strong belief in one God and by its adherents' experience of God's love. Christians are asked to reciprocate God's love by loving all others. According to Christianity, people have only one life to determine whether in the afterlife they will go to heaven or hell. Christianity as a whole emphasizes loving, compassionate actions and not committing sins. The ultimate of wellbeing in Christianity is to experience oneness with God along with the love of God and the love of others. (Tomer 2002, p. 35)

Islam is very much interested in establishing a society that is pure, that is free from the filth of sin and crime, a society in which men and women would live in total freedom from fear of all kinds, including fear of crimes and violence. It, therefore, wishes to build that society on foundations of piety, fear of God, and absolute justice, in which each person would respect the rights of other people and would not trespass on them. Islam has laid down universal fundamental rights for humanity that to be observed and respected under all circumstances. These rights can be realized in one's daily and social life as it provides both legal safeguards and a very effective moral system. In brief, whatever improves the wellbeing of an individual or a society is morally good, and whatever harms wellbeing is morally bad. A major goal of Islam is to provide mankind a practical and realistic system of life. It calls upon mankind not only to practice virtue but to establish it and to eradicate all that is harmful with the supremacy of God's conscience in all matters.

3.2.2 Subjective and Objective Wellbeing

General wellbeing comprises objective descriptions and subjective evaluations of physical, material, social, and emotional states together with the extent of personal development and purposeful activity, all weighted by a personal set of value. A definition of quality of life that disregards objective assessment of life conditions may not provide an adequate protection for the best interests of vulnerable and disadvantaged population. Expressions of satisfaction may simply reflect the difficulty of conditions commonly experienced by those with limited skills and little attachment to the mainstream society and its economy. For the objective assessment of Life domains we can use biological, material, social, behavioural, or psychological indicators. Subjective feelings about each area of life may be reflected in reports of satisfaction and wellbeing. (Felce and Perry 1995, p. 57- 62)

Subjective wellbeing is the response provided by one about the quality of his/her lives, which includes both cognitive judgments of life satisfaction and affective (positive and negative) evaluations of moods and emotions. Subjective wellbeing can reflect temporary influences, so it is moderately stable over time and across many situations. Then we can say that subjective wellbeing is a viable concept with long-term validity (Tomer 2002, p. 28). Neither ideal conditions nor perfect satisfaction can be arranged for or achieved by every member of a society or societal subgroup as the conditions and satisfaction with life inevitably vary from person to person and society to society. Aggregated data for a distinct group of interest may be match up to those for the total population to establish the situations either in their favour or in inconvenience. An acceptable quality of life requires the both expressed satisfaction with various aspects of life and objective descriptions of those aspects for the society as a whole (Felce & Perry 1995, p. 59)

Simply subjective wellbeing is one's perception about his/her life and the personnel satisfaction/dissatisfaction with the condition in which he/she lives. According to Tomer a person's wellbeing depends on the satisfaction of three basic kinds of needs: having (material and impersonal resources), relating (love, companionship and solidarity), and being (self-actualization and obverse of alienation). The relationship between the objective conditions and wellbeing is weak because the objective circumstances such as economic progress do not reliably satisfy all the above needs (Tomer 2002, p. 29-30).

Subjective wellbeing is not mutually exclusive with the objective condition prevailing in one's society and locality, but they are interrelated and have an effect on each other. 'Objective conditions like housing quality, level of pay, security on the street, economic stability and the services available are influence people's attitude of life satisfaction or happiness. Different people in different classes and inhabiting different places can react differently to the same objective conditions. This effect is measurable, with respect to the existing situation as well as the impact of some change that might result from public policy or private action' (Herbert and Smith 1979, p.15)

3.2.3 Measuring Wellbeing

Measuring wellbeing has four main issues. Should the focus be on objective indicators or subjective indicators of satisfaction, whether the data should collected from individuals or describe the functioning of systems, whether measures should be explicitly value-laden or value-neutral and the final issue is most apparent in the discussion of social indicator models, that is whether measures should be closely related to social policy and social change goals. For the collection of the data the method of objective measures of system functioning such as roles and social relations, income and consumption, and housing and safety, could be used. Individual-level measures could be employed in the form of subjective value-context measures like aspirations, expectations and distributive justice value or subjective wellbeing indices like life satisfaction, specific satisfaction and alienation (Raphael et al. 1996)

It is a very debatable topic that what should be given priority in the measurement. Several streams and methods are evolved a theoretical base for this, some relates and some differs. We can note the most important streams in this stream as population dynamics and urban ecology. Various factors like population dynamics, socio-political system, process of development, availability of resources and the existing level of living of people are affect the achievement of a desirable level of quality of life. Both objective and subjective dimensions of the quality of life are the functions of Population dynamics. Population dynamics affects all other factors and in turn is affected by them (Panda and Mishra 2001). Even though there are several variations in the methods, most of the modern urban ecologists believes that the only unique integrant to study urban social structure is urban ecology. Human spatial distribution in and around cities is the only major urban phenomena that remains sufficiently separable to provide a legitimate intellectual rationale. (Nottridge 1972, p.41)

3.2.4. Personal Capital and Social Capital

Personal capital and social capital are important in gaining insight regarding wellbeing. Social capital is anything that facilitates individual or collective action, generated by networks of relationships, reciprocity, trust, and social norms. According to Putnam 'physical capital refers to physical objects and human capital refers to the properties of individuals, social capital refers to connections among individuals, social networks and the norms of reciprocity and trustworthiness. In that sense social capital is closely related to what some have called "civic virtue." The difference is that "social capital" calls attention to the fact that civic virtue is most powerful when embedded in a sense network of reciprocal social relations. A society of many virtuous but isolated individuals is not necessarily rich in social capital' (Putnam 2000, p. 19)

Personal capital is a kind of human capital that relates to an individual's basic personal qualities and reflects the quality of an individual's psychological, physical, and spiritual functioning. A very important component of personal capital is the human capacity called emotional intelligence. The importance of personal capital is that as people grow, moving along the human development path, becoming more mature, more capable, and more conscious, a good part of this development can be characterized as personal capital formation. People are becoming more capable due to psychological or emotional, physical, and spiritual growth. These people are becoming more capable in all spheres of life, not just the work sphere and the personal sphere. If the investment in personal capital raises emotional intelligence, it may be because of greater social skills, increased motivation, or increased awareness and ability to regulate one's emotions that people have more and better beings and doings. (Tomer 2002, p. 36-37)

Social capital is productive, making possible the achievement of certain ends that in its absence would not be possible. When people are successful at transforming their relationships for the better, i.e., when these relationships become more loving, compassionate, and joyful, this involves a wellbeing increasing investment. Societal, cultural and personality characteristics such as social stratification, isolation, reciprocity, materialism, altruism, rivalry and envy have negative influence in well being (Coleman 1988, p. 98)

3.2.5 Social Distance and Social Differences

One's intimate sentiments and feeling may be shared only by a few or perhaps by none even the urban dweller is surrounded by crowded millions. The concept of social distance commonly employed to designate the degree of intimacy or understanding between individuals. It is highly important in analyzing social relationships in cities (Gist and Halbert 1954, p. 267).

Social differences do exists in towns. Patterns of physical separation and social differences linked in some way at least in larger towns. The kind and degree of segregation may fluctuate greatly but it will suggest a number of important considerations. It can see at community level rather than individual or personnel level. People organize in the basis of social status in terms of ethnicity, religion, profession and economy. 'People are separated because of social differences either voluntarily, compulsorily or accidentally and they want to remain separated at least in their housing' (Nottridge 1972, p.53).

Despite this fact the urban community is a well established, highly intricate social, ecological and economic organization in another perspective. Social processes are at work in urban areas to establish social control through assimilation in to new forms of community organization (Nottridge 1972, p.61). The land value, settlement type and even the presence of municipal aided amenities are outcome of the socio political approach and affluence of the residents of particular locality. 'The residential area of a city is distinctively a part of the city. The residential area, or address within the residential area, places a person within the context of a particular urban social structure'. (Beshers 1962)

The possibilities of urban environmental agendas are determined by the general conditions under which urban population are lived, rather than ecological rationale. Two fundamental shifts need to be taken in to account. First, urban environmentalism emerged and became institutionalized within the processes of neo-liberal urbanization and privatization .Second, both urban space and urban social life have become fragmented and individualized (Brand and Thomas 2005, p.58).

3.3 Urban Environment and Social Wellbeing

Urban environment and the wellbeing of the urban dwellers are related with each other in several perspectives. Quality of life and the quality of the environment are interdependent in ecological perspective. The wellbeing of individual and family are strongly bounded with the wellbeing of the whole ecosystem (Kathryn and Ronit 1999, p. 309). Quality of life is an outcome of relationship between people and their everyday urban environments. Understanding the nature of the person–environment relationship is the ideal geographical question that lies at the core of the sub-discipline of social geography. In the context of the built environment, this can be interpreted as a concern with the degree of agreement or disagreement between city dwellers and their urban surroundings (Pacione 2003b, p. 19)

Environmental quality was consistently argued as the prime requirement of competiveness and the ultimate measure of the everyday urban experience of citizens. A complex alliance of institutions ensured that environmental quality become the definitive feature of the urban development aspirations, urban management goals and urban evaluation criteria. The term environment includes all that which surrounds man: the natural, political, social elements, and the connection between them. It means the environment is the systematic reality which arises from the articulation of socially organized mankind with nature, through the process of development (Brand and Thomas 2005)

Healthy environment is important and essential part of quality of life. The objective environmental conditions, such as pollution levels, congestion may lead to illness and mental disorders under some disadvantageous conditions related to the individual characteristics of a person. Good air quality, low noise levels, clean and sufficient water, good urban design with sufficient and high-quality public and green spaces, an agreeable local climate or opportunities to adapt, and social equity are the environmental elements of a good quality of life. For biodiversity and population urban green infrastructure is very important. Urban ecosystems are artificial and providing specific habitats to the dwellers. It using the basic ecosystem services provided by nature and biodiversity for the survival and to deliver good quality of life. Nature and biodiversity originate from green areas within and outside cities (EEA 2010)

Human wellbeing is the extent to which individuals have the ability and the opportunity to live the kinds of lives they have reason to value, and is shaped by a wide range of instrumental freedoms. Human wellbeing encompasses personal and environmental security, access to materials for a good life, good health and good social relations, all of which are closely related to each other, and underlie the freedom to make choices and take action.

Security relates to personal and environmental security. It includes access to natural and other resources, and freedom from violence, crime and wars, as well as security from natural and human-caused disasters. Material needs relate to access to ecosystem goods and services. The material basis for a good life includes secure and adequate livelihoods, income and assets, enough food and clean water at all times, shelter, clothing, access to energy to keep warm and cool, and access to other goods which support better life. Health is a state of complete physical, mental and social wellbeing, and not merely the absence of disease or illness. Good health not only include being strong and feeling well, but also freedom from avoidable disease, a healthy physical environment, access to energy, safe water and clean air. Social relations refer to positive characteristics that define interactions among individuals, such as social cohesion, reciprocity, mutual respect, good gender and family relations, and the ability to help others and provide for children.

There are many other factors which partially important in the wellbeing of person and communities. Increasing the real opportunities that people have to improve their lives requires addressing all these components. This is closely linked to environmental quality and the sustainability of ecosystem services. Therefore, an assessment of the impact of the environment on individuals' wellbeing can be done by mapping the impact of the environment.

The environment in which the human beings live itself is the important factors contributing to social wellbeing. The shortage of an agreeable environment has an adverse effect on the development of the innovating ability, physical efficiency and productivity and thus on their earning capacity and their level of living (Ganguly and Gupta 1976, p. 108). Shelter and proper surroundings is considered as an important aspect of wellbeing of an individual and his family. Inadequate and improper housing affect adversely both health and productivity of habitants.... Medical facilities reflect either directly or indirectly the condition of health in general and hence the quality of life. Health is closely related to the nutritional and environmental aspects of living and even a comparatively slight but generally improvement in medical facilities may have a great impact on it (Ganguly and Gupta 1976, p. 65- 76).

The distinction between perceptions and evaluations of an environment and Subjective wellbeing is attained little attention of researchers. Persons may perceive particular qualities of their community. They may evaluate the quantity and quality of the public transportation facilities or the housing conditions of their community more or less favourably. They express their satisfaction or dissatisfaction with specific aspects of their community's quality of life and their personal wellbeing in the narrower sense (Walter-Bush 2000 pp. 2-3)

3.3.1 Social Area Analysis

Social Area Analysis is the technique for identifying segments within a city, or it is the process of identifying the urban sub-communities. The basic unit for social area analysis is the census tract, which is a group of contiguous blocks in a city, contains a definite number of people and intended to be a socially homogeneous as possible. Census information can be arranged in to three groups of economic, family, and ethnic characteristics. The index of economic status averages measures of rent, education and occupation. The index of family depends upon fertility ratio, women not in the labour force and ratio of single family dwellings. Finally the index of ethnic status can take from information on race, nativity, and surnames (Reissman 1964, pp. 86-87).

Social area analysis is a somewhat different scheme for classification. It contains its own particular advantages and disadvantages and like most other classification devices it has the arbitrary manner of defining categories, and lack of relevance for a broad theory. It also has the advantage of some statistical precision and objective measurement as well as the ability to compress a great deal of data in to a single composite score (Reissman 1964, p. 89).

3.3.2 Social Indicators Movement

The insufficiency of information relating to social wellbeing in academic and government circle in 1970s had led to what may be referred as the social indicators movement. As the social conditions are vary in time and space the development of social indicators involves the measurement of them. The healthy information about the nature and performance of the social system as well as the economic system is the intention of this movement. 'Social indicators should ideally measures the state of and changes over time in major aspects of dimensions of social condition that can be judged normatively, as part of a comprehensive and inter related set of such measures embedded in a social model and their compilation and use should be related to public policy goals'. (Smith 1973, p. 54)

According to Knox Social indicators is 'an aggregate or composite measures of wellbeing, or of some element of it, and are generally designed to facilitate concise and comprehensive judgement about levels of social welfare. In most cases they are aimed at improving information systems for decision- making: to assess what is happening, to pave the way for policy decisions, and to monitor the effect of policies'. The concepts of level of living, standard of living, social wellbeing, social welfare and level of satisfaction all symbolize the same general notion of wellbeing as the quality of life, but all may have different connotation or implications (Knox, 1975 pp 8-10)

The modern social indicators movement can be seen as a new approach to a very old and important concern through empirical social science concepts and methods. The modern social indicators movement get under way in the mid 1960s and has been in energetic worldwide development for the past fifteen years. The failure of traditional economic statistics do to provide the broad-based information needed for monitoring social change, evaluating social programs, guiding policy development, and in general assessing quality of life or levels of wellbeing had bring forth the social indicators movement. Significant efforts have been devoted by social scientists, statisticians, government administrators, and others to laying the conceptual and operational foundations on which more broadly based social information systems can be built. At the national level, many of the more developed countries and also some of the developing countries have published some kind of social indicators report (Andrews 1983 pp.215-216)

The objective living-conditions and subjective satisfaction with them is the primary field of application of the social indicators concept of wellbeing, medical concept of quality of life emphasizes the health related subjective wellbeing of the individual. The social indicators concept is concerned both with objective resources and with their subjective perception, which make tensions between objective and subjective indicators. The concept was designed to have an immediate political impact and to be relevant to social planning, for which objective indicators such as quality of living conditions, environmental quality, quality of work conditions, of social services and of health care were mandatory. The subjectivist notion of quality of life is largely autonomous of changing social and cultural values. If the quality of life is analysed by objective characteristics, it's applicability to specific cultural norms and ideals will deteriorates. Quality of life must focus on subjective wellbeing rather than on the nature of the objective conditions on which subjective wellbeing depends. (Dieter 1999, pp. 28-30)

Quality of life is something related to subject's own evaluation of his personal state. Quality of life, then, is more closely connected with preferences or attitudes than with the quality of subjective states judged from an impersonal point of view. 'Quality of life cannot depend on how an individual evaluates certain objective events but only on how he evaluates his subjective states resulting from the event on the occasion of its happening. Equally, quality of life does not depend on how a future subjective state is evaluated before its occurrence but on how it is evaluated when it actually occurs'. (Dieter 1999 pp.31)

3.3.3 Territorial Social Indicators

Territorial social indicators are a necessary and logical extension of any realistic system of social reporting. People locally experience the prosperity stresses expectation and satisfaction of their locality. National social indicators are aggregates of these conditions and as such may cover important problems at local level (Knox 1975, p.11). A definition of social wellbeing can be set as the compilation of the data necessary to determine the degree and regularity of spatial variation, and to set down these observations in numerical and cartographic form. The outcome should be a set of "territorial social indicators" (Smith 1973, p. 7-8).

Two distinct types of social indicators are suitable for measuring societal and individual wellbeing. First one is the objective indicators that describe the environment within which people live and work. It includes the levels of health care provision, crime, education, leisure facilities and housing. The second one is subjective indicators aimed to describe the ways in which people perceive and evaluate conditions around them. (Pacione 2003a, p. 21)

Objective indicators are the commodities in human environments that promote to the good life and that represent the average conditions of a large population. Objective indicators are representing the conditions of people's lives and the environment in which they live, but not the subjective evaluation of a person whose life is being evaluated. The objective indicators usually refer to three aspects of public welfare: economic, physical/environmental, and social (Kathryn and Ronit 1999, pp. 309-310)

Subjective indicator (Perceptual indicators) of life quality attempt to quantify the experience of life rather than the conditions of life, and the individual is the unit of analysis. The studies using perceptual indicators concentrate on the definition of human needs and the satisfactory fulfilment of these needs. The researchers who employ the subjective indicators refer "wellbeing", rather than "welfare" as the unit of observation is not a large population and their conditions of life, but an individual and his/her experiences of life. Family wellbeing can be denoted by an individual's perceptions of the extent to which his/her material and emotional needs are satisfied. (Retting and Ronit 1999, pp. 310)

U.S. Department of Health, Education, and Welfare in 1969 defined social indicator as a statistic of direct normative interest which facilitates concise, comprehensive and balanced judgments about the conditions of major aspects of a society. It is in all cases a direct measure of welfare and is subjected to the interpretation that, if it changes in the correct direction, while other things remain equal, things have gotten better or people are "better off". Thus, statistics on the number of doctors or policemen could not be social indicators, whereas figures on health or crime rate could be.

Socioeconomic welfare, including population (composition, growth and distribution), labour force and employment, income, knowledge and technology, education, health, leisure, public safety and legal system, housing, transportation, physical environment, social mobility and stratification are the major categories of a social report using indicator systems. Social participation and isolation could also be assessed with target upon: family, religion, politics, voluntary associations, and alienation. (Raphael et al. 1996, p.76)

Geographers have initiated the concept of territorial social indicators to identify and analyse socio-spatial variations in quality of life at different geographic scales, varying from global to local. The objective measures of quality of life, derived from primary field surveys have been employed in most of the researches using territorial social indicators. This line of research has contributed valuable insights into such questions as the extent and distribution of substandard housing, and the differential incidence of deprivation within the city (Pacione 2003, p.20).

Urban liveability is a relative rather than absolute term whose precise meaning relies on the place, time and purpose of the assessment, and on the value system of the assessor. This view contends that quality is not an attribute inherent in the environment but is a behaviour-related function of the interaction of environmental and personal characteristics. In order to obtain a proper understanding of urban environmental quality it is essential to employ both objective and subjective evaluations. In other words, we must consider both the city on the ground and the city in the mind (Pacione 2003b, p.20).

UNO put forwarded a standard measure for level of living using the component of health including demographic conditions, food and nutrition, education including literacy and skills, conditions of work, employment situation, aggregate consumption and savings, transportation, housing including household facilities, clothing, recreation and entertainment, social security and human freedom. Drewnowski also established a level of living index using the components nutrition, clothing, shelter, health, education, leisure, security, social environment and physical environment and Smith bring about an index for social wellbeing in United States using income, wealth and employment, the living environment, health, education, social order, social belonging, recreation and leisure. There are many other indices put forwarded by several scholars, but owing to the different setting of site and manner of problem the investigators must be rearrange or establish the indicators for unbiased and relevant results.

4

HISTORICO-GEOGRAPHICAL SETTINGS OF CALICUT CITY

Kerala ranks highest in India regarding social development indices such as elimination of poverty, primary education and healthcare. Kerala has the highest literacy rate (91%), life expectancy (73 yrs) and sex ratio (1058) amongst the all states of India. UNICEF and the World Health Organization (WHO) designated Kerala the world's first "baby-friendly state". The state is also known for the traditional medicinal system Ayurveda which attract increasing numbers of medical tourists from the country as well as overseas. It is a place that offers real hope for the future of the Third World and can compare developed nations in infant mortality rate, literacy rate and its birth-rate is below America's and falling faster. Kerala's residents live nearly as long as Americans or Europeans.

Kerala is home to 2.76% of India's people, and have a density of 819 persons per km^2. However, Kerala's population growth rate is far lower than the national average. As per the Census 2001 data, Hindus, who constitute 56.20 per cent of the total population, is the most prominent religious community in the state. This is followed by Muslims who form 24.7 per cent and Christians who constitute 19 percent of the total population.

The colonial era in the history of Kerala is started with the arrival of the Europeans after Vasco da Gama reached Kappad in 1498 near Kozhikode and was followed by the arrival of a number of Europeans from Portugal itself and other countries. The main aim of their visit was trade and discovery of a shorter sea route to the Malabar Coast, but the prevailing political instability among the small provinces granted them entry into the administration. They used the rivalry between the provincial rulers, and provided military assistance to one ruler against the other. The first Europeans who established a stronghold in Kerala were Portuguese followed by the Dutch and the British. A number of battles were fought between the provincial rulers against each other and against the Portuguese and finally in 1524 Vasco da Gama was appointed as the Portuguese Viceroy of Kerala. Kochi and Kozhikode were the main provinces of that time. Zamorins, the rulers of Kozhikode, fought a number of battles against the Portuguese to fade away them from their land.

After Portuguese, the Dutch reached Kerala and established the Dutch East India Company in the year 1592. Dutch army arrived at the Malabar coast in 1604 and entered the arena of Kerala politics by utilizing the rivalry between Kochi and Kozhikode. The Dutch supremacy lasted only for a short period before the British entry into Kerala. British supremacy in Kerala started after four Anglo Mysoor war with Hyder Ali and Tippu Sultan in the mid seventeenth century and lasted for the next 200 years till independence. These rulers especially the British led many changes in the social and cultural life of Kerala.

4.1 Physical Setting of Calicut District

Kozhikode, one of the fourteen districts of Kerala State is situated on the south west cost of Indian sub continent and located on the northern part of the state Kerala. The district is bounded in the north by Kannur district, east by Wayanad, south by Malappuram district and West by the Arabian Sea. The geographical area of the district is 2344 sq kms. It lays between 11°08' and 11°50' North latitudes and 75°30' and 76°08' East Longitudes. The district has a coastal length of 71 kms. Topographically the district has three distinct regions - the sandy coastal region including sand bars and marshes, the rocky highlands formed by the hilly portion of the Western Ghats and lateritic midland with undulating topography. From the total area of 2344 sq.kms, the sandy coastal belt contributes about 15.5 per cent, lateritic midlands about 57.3 per cent and rocky highlands about 27.2 per cent.

The district is falling under the tropical humid climatic region. During December to march, practically no rain is received, and from October onwards the temperature gradually increases to reach the maximum in May, which is the hottest month of the year. The highest maximum temperature recorded at Kozhikode was 39.4 °C during the month of March 1975 and lowest temperature was 14 °C recorded on 26th December 1975. Humidity is very high all along the coastal region. The district has a generally humid climate with a very hot season extending from March to May. The rainy season is the South West Monsoon, which sets in the first week of June and extends up to September. The North East Monsoon extends from the second half of October through November. The average annual rainfall is 3266 mm. Kozhikode district has six important rivers, namely Mahe, Kuttiady, Kora, Kallai, Chaliyar and

Kadalundi. All these rivers originated from the eastern part of the district and flow to western side to join the Arabian Sea. The whole drainage area of the Kallai river, Kora river and Kuttiady river are within Kozhikode district.

The History of the district is inevitably intertwined with the history of the city of Calicut. From the very beginning of the documented history Kozhikode was under control of the rulers known as Zamorins. The conflict is begun with the Portuguese invaders after a long period of bilateral trade with them since the arrival of Vasco-Da Gamma in 1498 at Kozhikode. The Zamorins later allied with the Dutch to weaken the Portuguese and by the mid-17th century the Dutch had captured the Malabar Coast spice trade from the Portuguese. In 1766 Hyder Ali of Mysore captured Kozhikode and much of the northern Malabar Coast, and came into conflict with the British based in Madras presidency, which resulted in four Anglo-Mysore Wars.

However present day Kozhikode District was among the territories abandoned to the British East India Company by Tipu Sultan of Mysore in 1792, at the conclusion of the Third Anglo-Mysore War. The newly-acquired British possessions on the Malabar Coast were organized into Malabar District, which included present-day districts of Kannur, Kozhikode, Malappuram, Palakkad, and Wayanad. Calicut served as the administrative headquarters of the Malabar district. After India's Independence in 1947, Madras Presidency was renamed Madras State. When Madras state was divided along linguistic lines by the State Names Reorganization Act, Malabar District was combined with the erstwhile state of Travancore-Cochin and Kasaragod District to form the state of Kerala on 1 November 1956.

The history of Kozhikode district as an administrative unit begins from January 1957. When the states of the Indian Union were reorganized on linguistic basis on 1st November, 1956, the erstwhile Malabar district was separated from Madras state (Tamil Nadu) and added to the new unilingual state of Kerala. The state government ordered the formation of Malabar in to three districts. The Kozhikode district thus came into existence on 1st Jnauary 1957, originally consisting of five *taluks*, they are, Vadakara, Koyilandy, Kozhikode, Ernad & Tirur. With the formation of Malapuram district on 1st June 1969 & Wayandu on 1st November 1980, Kozhikode district now consist of one revenue division, three *taluks* (Kozhikodu, Vadakara and Koyilandi) twelve blocks, 78 *panchayats* and 117 villages.

The Kozhikode district is one of the industrially advanced areas of the state, with many small scale industries flourishing from early days. The timber industry has great influence on the economy of this district. There are 1,564 registered forest based industrial units in the district; all these units are concentrated at Kallai, Cherunavvur and Feroke area. The finished timber goods are marketed locally and are also exported. Another major industry is the tile industry. The tile factories are mostly concentrated at Feroke and Cheruvannur area, which is rich in quality clay deposits due to lack of enough demand and lack of labours most of them are closed. A number of traditional cottage industries are there in the district, the important among them are coir and coir products, weaving, pottery, and oil industries. Apart from the above given picture, there is a ship breaking unit and a boat building yard under public sector at Beypore.

According to the 2011 census Kozhikode district have a population of 3,089,543, and population density of 1,318 inhabitants per square kilometre, sex ratio of 1097, and literacy rate of 95.24%. Its population growth rate over the decade 2001-2011 was 7.31 %. Hindus constitute the majority of the population and followed by the Muslim and the Christian communities respectively. The entire district is extensively covered with wide network of roads and well served railway lines. The total length of road network comes to 2051.387 kms, it includes the national highways, state highways, district roads, village roads and city roads. The railway line runs through the costal side of the district with the length of 79.5 kms, broad-gauge line and 17 stations. Kozhikode has an airport with domestic and international flights. The district has two intermediate ports at Kozhikode and Beypore and three fisheries harbours at Puthiyappa, Beypore and Chombal.

4.2 Geographical Setting of Calicut City

Calicut is a medium sized coastal city and one of the main commercial centers of Kerala which serves whole the northern part of the state i.e. Malabar region. The city was founded on a marshy tract along the Arabian coast in 1034 A.D. It is one of the five Municipal Corporations in Kerala. It became a Municipality on 3 July 1866 with a population 36,602 and inhabiting in an area of 28.48 sq. kms. It was later made a Municipal Corporation in 1962. Currently, the Corporation is spread over an area of 84.232 sq. kms. with a population of 432,097 (2011)

Traditionally Calicut was a famous port for trading and has a long time relationship with Arabian Peninsula. Foreigners called it by different names, for Arabs it was Kalikut, for Chinese it was Kalifo and in local language it is Kozhikode. Calicut is the anglicized form often called by Europeans. Kozhikode also known as Calicut is the third largest city in the southern state of Kerala in India. It is the headquarters of Kozhikode district, and was formerly the capital of an independent kingdom, and later of the erstwhile Malabar District. This city is famous as the place where Vasco da Gama, the first European to sail directly from Europe (from Lisbon, Portugal) to India, landed in 1498.

Calicut has witnessed a building boom in recent years particularly there is a boom in the number of malls, multi storied flats and apartments built in recent years. According to data compiled by economics research firm Indicus Analytics on residences, earnings and investments based on six parameters – health, education, environment, safety, public facilities and entertainment, Kozhikode ranked as the second best city in India to reside in. Kozhikode was ranked eleventh among Tier-II Indian cities in job creation by a study conducted by association of chamber of commerce Delhi in 2007. Kozhikode was declared the first litter- free city in India in 2004. A 'Hunger-Free Kozhikode' project was initiated in January 2009 following which Kozhikode was declared the country's first hunger-free city. Kozhikode is come under the radar of the IT industry with the development of Cyber Park by the Kerala government. This is the third IT 'Hub' in the state developed on the lines of Thiruvananthapuram Techno Park and Kochi Info Park.

4.2.1 Location

Calicut city is located at 11°15' N and 75°47' E in the Arabian coast of southern state Kerala, India. It has an elevation of 1 metre (3 feet) along the coast with the city's

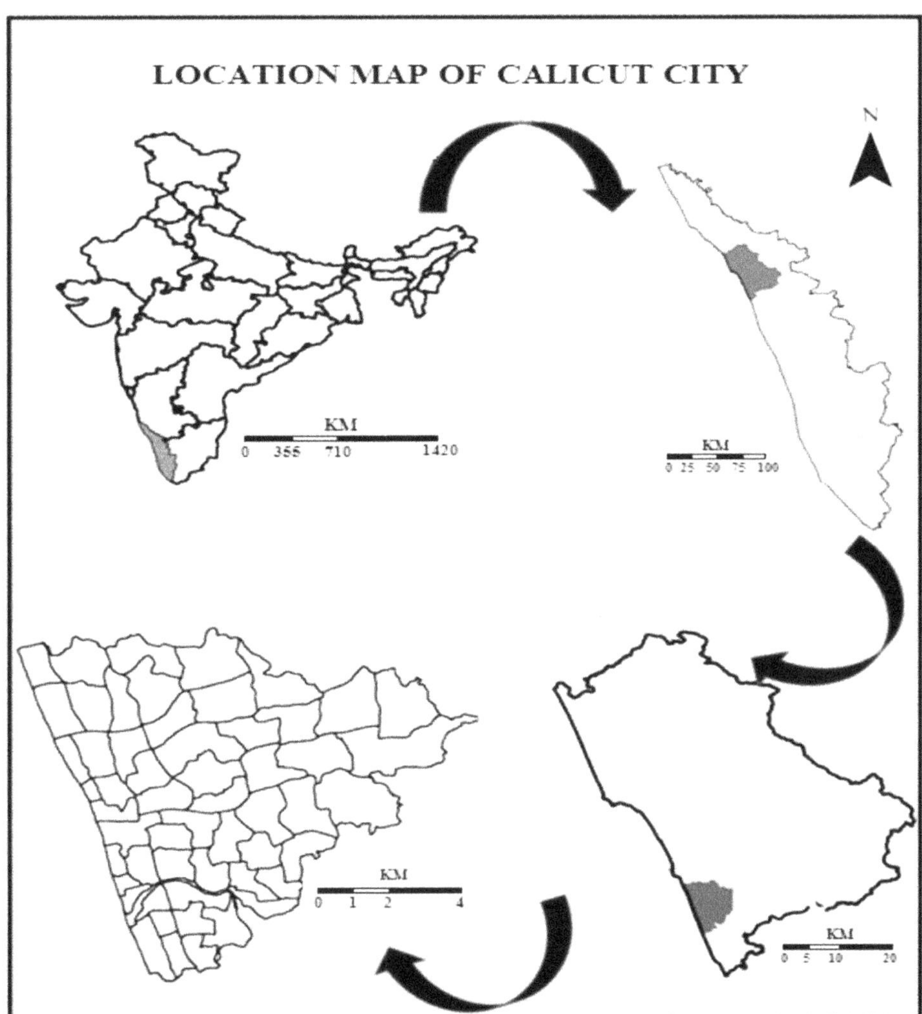

Figure: 4.1 Location Map of Calicut City

eastern edges rising to at least 15 metre, with a sandy coastal belt and a lateritic midland. The city has a 15 km long shore line and small hills dot the terrain in the eastern and central regions. It serves whole the northern Kerala especially known as Malabar region. The city is about 450 km far from the capital of Kerala i.e. Trivandrum and about 200 km from the largest city of Kerala i.e. Ernakulum urban agglomeration. It has been an icon in the history and politics of Kerala due to its strategic position.

4.2.2 Geography

The Geographical outlook of city area and surrounding areas are almost similar to the other parts of the district comprising coastal and midland zones in the typical classification of land in Kerala as low, mid and high lands. Lagoons and backwaters characterize the low land which receives drainage from the rivers and tidal water from the sea. The coastal plains exhibit more or less flat, narrow terrain with landforms such as beach ridges, sandbars, backwater marshes, etc. The lowland is often subjected to salinity intrusion. Moving from the sea to the east, the surface amass into slopes and clustered and isolated hills with numerous valleys in between them formed due to floods and sediment transportation especially by fluvial cycle. The soil conditions and climate are typical for cultivation of different spices, coconut and areca nut and normal for other crops like vegetables and flowers. Garden lands form major share of land used for cultivation with cash crops and oil crops.

4.2.3 Soil

The Soil group of Calicut city can be divided into 4 major types – The coastal alluvial soil along the coastal plain and in low lying areas, Riverine alluvial soil along the River banks, Red loam soil and brown hydromorphic soil. The coastal zone is covered by Cenozoic alluvium sediments of recent age at few places the crystalline rocks are cross-cutter by basic dykes, comprises mainly of charnockites with enclaves of mafic granulites belongs to Lower Precambrian age. In the midlands at places, these rocks are covered by laterites. The surfacial geological feature s of the coastal belt mainly consists of sand dunes- occasionally interspersed with sandstones and clays.

The Soil conditions are very good for cultivation of spices and coconut especially and normal for other crops. Garden lands form major share of land used for cultivation. In the city nearly 5500 Hectares of land is being used for cultivation and nearly 321 Hectares are waterlogged area.

4.2.4 Climate

The city has a generally tropical humid climate with a very hot season and wet season or rainy season. According to the Koppen's climate classification it comes under tropical monsoon climate. The hot season extends from March to May without rain except a short spell of pre-monsoon Mango Showers. The main rainy season is from June to September and the rain receives from the south west monsoon which accounts for about 62% of rainfall received by the region. Some rain is also receives by the north east monsoon in the months of October and November. It accounts for 14% of rainfall and the rest of rain fall occur in winter season from December to January. The average annual rainfall is 3,084 mm. winters is seldom cold and skies are clear and air is crisp. The nearness to sea is not allows much variation in temperature and the average temperature is about 28°c.The highest temperature ever recorded in the area was 39.4°c in March 1975. The lowest was 14°c recorded on 26 December 1975.

Table: 4.1 Climatological Table of Calicut (1971-2000)

Month	MeanTemperature(°C)		MeanTota lRainfall(mm)	Mean Number of Days With	
	Daily Minimum	Daily Maximum		Rain	Thunder
Jan	22.2	31.8	1.6	0.1	0.3
Feb	23.5	32.2	1.7	0.2	0.2
Mar	25.1	32.9	14.2	0.7	1.2
Apr	26.2	33.2	76.6	3.6	6.8
May	25.9	32.6	217.8	8.3	9.6
Jun	24.1	29.6	837.6	23.1	5.9
Jul	23.5	28.5	816.5	24.5	2.8
Aug	23.6	28.5	454.8	20.5	0.8
Sep	24.1	29.8	222.4	10.6	0.3
Oct	24.0	30.7	262.8	10.7	9.5
Nov	23.7	31.4	149.4	6.2	6.2
Dec	22.7	31.8	28.4	1.6	1.6
Annual	**24.1**	**31.1**	**3083.9**	**110.2**	**48.0**

Source: Indian Meteorological Department.

4.2.5 Rivers and Canals

There are large number of Rivers and Lakes in Calicut district. Many rivers originating from the Western Ghats run along the outer reaches of the Calicut city. These include the Chaliyar river, Kallayi river, Kora river and Poonoor river. Of these, Kallai river that runs through the southern part of the city has been the most important culturally and historically for the city. Apart from it there are many tributaries of these rivers spread all across the city.

The Canoly Canal running across the heart of the city has a length of 11 km. The width of the canal ranges from 6 m to 20 m. The depth of water in the peak summer varies from 0.5 m to 2 m. Canoly Canal was built in 1848 to connect the Korapuzha river in the north to Kallayi river in the south. It functions as a drain to reduce flooding in the city during the rainy season and as a navigation channel.

4.2.6 Wetlands

A deep network of lakes, canals, estuaries and wetlands runs through the city. Prominent among these is the Canoli Canal built in 1848 to connect the Korapuzha River in the north to Kallayi River in the south. A vast system of wetland (mangrove) forests pervades the city from Kallai River in south to Eranjikkal in the north forming the most crucial lifeline of the city. The Kotooli wetlands are notable in this respect. Kottoli Urban Wetlands spread over 200 acres, is located in the heart of the city between Eranjipalam and Baby Memorial Hospital Junction. This is one of the 27 wetlands of natural importance identified by the Government of India for conservation under National Wetland conservation programme.In spite of the wetland's crucial role in maintenance of biodiversity, climatology balance, and ground water table maintenance and in flood management, they have come under increasing threat due to unprecedented management and self-serving commercial interests, compounded by an ignorance of their significance by the local population.

4.2.7 Land Use

Land use is the method or system of land developed and used in terms of the types of activities allowed for agriculture, residences, industries, transportation etc. and the size of buildings and structures permitted, or the arrangements, activities and inputs people undertake in a certain land cover type to produce, change or maintain it. Land use varies from area to area. In rural areas land use can include forestry and farming. In urban areas land use could be housing or industry. Urban land use models attempt to simplify the way land is used in urban areas.

The rapid growth of urbanization influenced the expansion and constant change of urban land use. Physical features, social condition and social organization, political power and economical factors have played their decisive roles in forming land use pattern of any town or city. Calicut is mainly residential and commercial city sharing more than half of its geographical area to the residential purpose. Other important share can be noticed of public and semi public area, water courses and the parks and open spaces followed by the industrial areas and agricultural lands.

Table: 4.2 Existing Land Use in Calicut

Land use	*Area in sq. km*	*Land use (%)*
Residential area	44.03	52.27
Commercial area	1.22	1.45
Industrial area	3.79	4.50
Public & semi public area	14.28	16.95
Transportation area	0.64	0.76
Agriculture	1.87	2.21
Water course	11.17	13.26
Parks & open spaces	7.24	8.60
Total	**84.24**	**100.00**

Source: KSPCB- Model SWM System- Kozhikode, 2006.

4.3 Historical Background

4.3.1 Evolution

It can be said that the history of Kerala or then Malabar at least from 12th century A.D. onwards to the first decades of 16th century is nothing but the history of Calicut. Calicut was known by various names among the foreigners and other merchants such as Kalikooth for Arabs, Kallikkottai for Tamils and Kozhikode for Malayalees. It seems that from the name of fort at Kallai that the name of Kallikkottai in Tamil, Kalikuth in Arabic and Calicut in European languages originated. In fact there was no city of Calicut before 12th century of Christian era. Calicut starts figuring in the political history of Malabar with the disintegration of Perumal kingdom, therefore, the ups and downs in the future of Calicut were closely interwoven with those of Zamorins who supersede the Perumal kingdom or so called Chera dynasty in 1122 A.D.

The most prominent factor that helped the rise of Calicut from 12th century was the efforts of the Arabs and the native Muslims who were their descendants and followers. At the same time the fore-sighted political leadership of the *Samuthiris* and their advisors provided the necessary background for these efforts. Thus a *Nagaram* (market place) was built, and the place is became the main market place. Calicut has a long history of trade with the Arabs and Chinese, and it made the city a popular trading centre. The Historian Prof. K.V. Krishna Iyer claimed that the city was founded on a marshy tract along the Arabian coast in 1034 A.D. following the collapse of the powerful *Chera* Kingdom. As Calicut offered full freedom and security, the Arab and the Chinese merchants preferred it to all other ports. The globe trotter Ibn Battuta (A.D. 1342-47) notes: "We came next to Calicut, one of the great ports of the district of Malabar, and in which merchants of all parts of the world are found".

Vasco da Gama landed at Kappad near the city in May1498, as the leader of a trade mission from Portugal and was received by the Zamorin himself. In 1503 a Portuguese trading post was built in Chaliyam on the mouth of the river Chaliyar. In 1766 Hyder Ali of Mysore captured Calicut and much of the northern Malabar Coast, and came into conflict with the British based in Madras, which resulted in four Anglo-Mysore Wars and it came finally under the British rule.

4.3.2 Early Calicut in Foreign Accounts

Remarkable accounts of the city and surrounding areas and the conditions prevailing then can be compiled from the narratives of various travellers who visited the port city of Calicut. Ibn Battuta (1342–1347), a six times visitor in the Calicut, gives us the earliest preview of life in the city as "one of the great ports of the district of Malabar where merchants of all parts of the world are found".

The Chinese sailor Ma Huang (1403 A.D), glorify the city as a great emporium of trade by merchants from around the world. He makes note the unique system of calculation by the merchants using their fingers and toes and the matrilineal system of succession practiced in the city. The ambassador of Persian Emperor Sha-Rohk, Abdur Razzak (1442–43) explains the city as harbour offering perfect security and perceives precious commodities from several maritime countries especially from Abyssinia, Zirbad and Zanzibar.

The Italian Niccolò de' Conti (1445), may be the first Christian traveller who reached Calicut from Europe. He describes the city as the market place of pepper, ginger, a larger kind of Cinnamon, myrobalans and Zedary. He calls it a noble emporium for all India. The Russian traveller Afanasy Nikitin (1468–74) calls 'Calecut' a port for the whole Indian sea and describes Calicut as having a "big bazaar" (which exists even now as a market place). Many other travellers visited Calicut other than who came for trade and colonization purpose and gave a sound position and explanation for Calicut in their travel accounts.

4.3.3 Colonial Period

The ports of the Malabar Coast have participated in the Indian Ocean trade in spices, silk, and other goods for over two millennia. There are documented visits in as early as the 14th century, by Chinese travellers such as Zheng He. Kozhikode had emerged as the centre of an independent kingdom by the 14th century, whose ruler was known as the *Samoothirippadu* (Zamorin). Arrival of Portuguese led by Vasco da Gama established the colonial presence in city and it continued by Dutch and finally British up to 1947.

The Portuguese period

The colonial presence in Calicut city is started from the arrival of Portuguese in Calicut in May 1498 led by Vasco da Gama and it continued for a long period. During the 16th century the Portuguese set up trading posts from north in Kannur to the south in Kochi, but the Zamorin resisted the establishment of a permanent Portuguese presence in the city, although in 1509 the kingdom was forced to accept a Portuguese trading post in Chaliyam. The fort was used by the Portuguese to attack Zamorin's interests. Later on because of the problems arose between Zamorin and Portuguese Calicut lost its prominence in 16th century and the centre trading activity is shifted to Cochin but by the beginning of the 17th century Calicut regained its glory and continued it for a long period. Due to the continuous threats and long battles with locals and later with Dutch the Portuguese cannot give much care in the development of city. Moreover their influence was restricted only in those places which handed over to them by Zamorines, out of the whole city, and after all their main concern was on the trade of spices and other commodities, than the involvement in administration.

The Dutch period

Dutch voyagers led by Steven van der Hagen arrived in Calicut in November 1604 and marked the beginning of the Dutch presence on the Indian coast. The Dutch had a more favourable relation with the Calicut and were provided greater participation in the ongoing trade. The Zamorins later allied with Portuguese's rivals, the Dutch, and by the mid-17th century the Dutch had captured the Malabar Coast spice trade from the Portuguese. Dutch people also couldn't contribute to the city development because they are in confronts or rivalry with the Portuguese for a long period.

The British period

The British reached Calicut in 1615, led by Captain William Keeling. In 1766 Hyder Ali of Mysore captured Calicut and much of northern Malabar Coast, and came into conflict with the British based in Madras, which resulted in four Anglo-Mysore Wars. Kozhikode and the surrounding districts were among the territories surrendered to the British by Tipu Sultan of Mysore at the conclusion of the Third Anglo-Mysore War in 1792. The newly acquired possessions on the Malabar Coast were organized into Malabar District of Madras Presidency, and Calicut became the district capital. Calicut remained the headquarters of the Malabar district under Madras state.

British managed to contribute much in the city development of Calicut because they had complete power of the administration of the city, after losing the power from Tippu sultan and Zamorins. They used Calicut city to administrate whole the Malabar region and they developed several other small settlements and bungalows like Nilambur, Kannur and Wayanadu. The most important city development that done by British, is the construction of Canoly canal in the year 1848 through the heart of the city for the navigation, flood control, and as a sewage drain which prevail even now without much distortion. They also set up a military barrack at Westhill area in 1849. Calicut city is given the status of municipality by British itself in 1866. And in 1872 a mental hospital established in Kuithiravattom, on which the cavalry of Zamorines were present. In the British period the Kallai was the second largest timber and furniture exporting port in Asia. The railway line, railway station, many important roads, bridges educational institutions and hospitals are also of their contribution which enhanced the growth of both physical and social environment.

4.3.4 Post Independence Period

After Indian Independence in 1947, Madras Presidency became Madras State. In 1956 the Indian states were reorganized along linguistic lines, and Malabar District was combined with the state of Travancore-Cochin into the new state of Kerala on November 1, 1956. Malabar District was divided into the districts of Kannur, Kozhikode, and Palakkad on 01 January 1957 and Calicut became the capital of Kozhikode district. It was a municipality since 3 July 1866, established with a population of 36,602 and inhabiting area of 28.48 sq. kms. After independence in 1962 it became a Municipal Corporation. Currently, the Corporation is spread over an area of 84.232 sq. kms with a population of 436556 (2001) and is divided in to 51 municipal wards and 39 revenue wards. The actual growth of the city is started in this period based on the capital infrastructure left by the British East India Company, and following their developmental model with indigenous architecture, engineering and planning.

4.4 Administration

Calicut is one of the five Municipal Corporations in Kerala. It became a municipality on 3 July 1866 with a population of 36,602 and inhabiting in an area of 28.48 sq. kms. It later made a Municipal Corporation in 1962. Currently, the Corporation is spread over an area of 84.232 sq. kms with a population of 436556 (2001). For administrative propose the city is divided in to 51 Municipal Electoral Wards and 39 Revenue Wards. The Corporation Council which constitute of elected representatives from each electoral wards control and coordinate the development programs. The election process is totally democratic. Mayor elected from these representatives that is councillors from each wards along with executive council lead the planning and allocations.

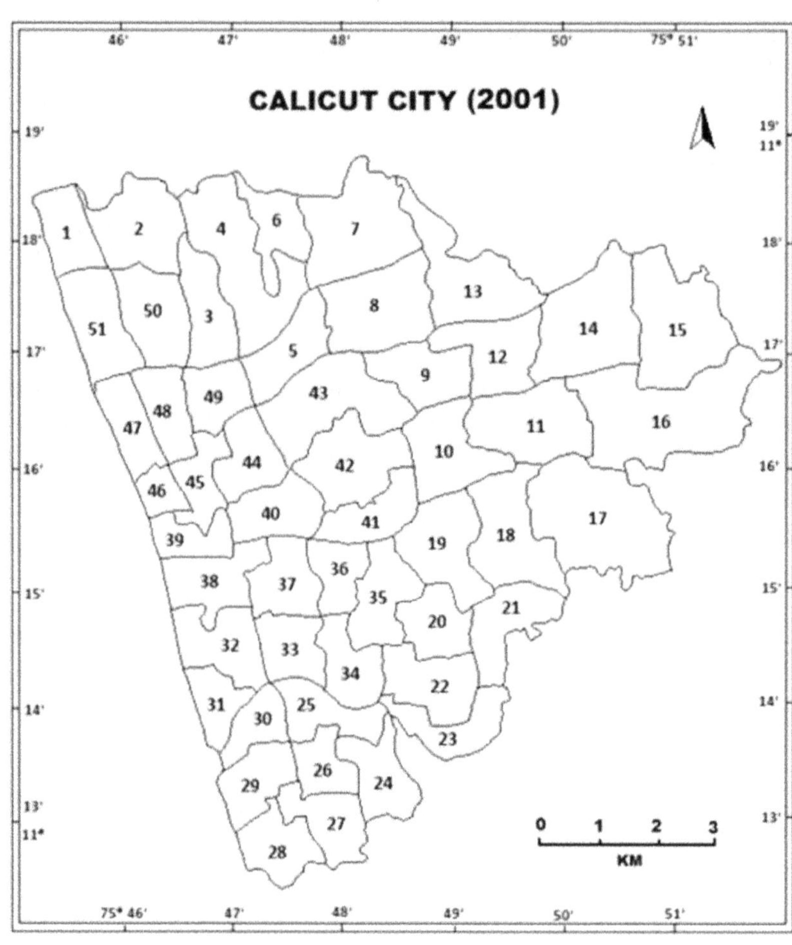

Figure: 4.2 Calicut City(2001)

1. New Bazar
2. Edakkat
3. Easthill
4. Kuruvisseri
5. Malaparamba
6. Vengeri
7. Kannadikkal
8. Paroppadi
9. Chevarambalam
10. Kudilthodu
11. Cheveyur
12. Silverhills
13. Poolakkadavu
14. Moozhikkal
15. Chelavoor
16. Mayanadu
17. Kovoor
18. Nellikkode
19. potamal
20. Kommeri
21. Pokkunnu
22. Mankavu
23. Kinasseri
24. Thiruvannur
25. Kallai
26. Panniyankara
27. Meenchanda
28. Koya Valappu
29. Payyanakkal
30. Chakumkadavu
31. Pallikkandi
32. Idiangara
33. Chalappuram
34. Azhchavattam
35. Kuthiravattam
36. Puthiyara
37. Palayam
38. Big Bazar
39. Vellayil South
40. Thiruthiyad
41. Kotooli South
42. Kotooli North
43. Civil Station
44. Eranchippalam
45. Karaparamba
46. Nadakkavu
47. Vellayil North
48. Thoppayil
49. Chakorat Kulam
50. Westhill
51. Varaykal

4.5 Infrastructure

Roads and other means of transportation lines play an important role in the economic development of any place. So indices of per capita availability of roads and level of connectivity would have better indicators of urban social wellbeing.

4.5.1 Transport and Communication

Almost all parts of the city and outer zone are well connected with communication facilities. The density of transport and communication is quite well developed in the district and the city particularly has a well- developed infrastructure for intra- city, inter- city, national and international travel. The city has three bus stands namely Palayam, KSRTC and Mofusil bustands.

Table: 4.3 Road Network in Calicut City

Road Category	*Length (Km)*	*% Share*
Black Top Road	361	49.93%
Metal Road	54	7.47%
Concrete Road	49	6.78%
Others	259	35.82%
Total	**723**	**100.00%**

Source: Directorate of Urban Affairs, Govt. of Kerala.

Kozhikode city has a road network of 723 kms within the total area about 83sq.km of which nearly 57% is surfaced with a road density of 8.7 km per sq. km. Mavoor Road; Mini Bypass from Meenchanda to West Hill; Beach Road; CH Flyover; Oyitti Road; Town Hall Road; Railway Station Road; Gandhi Road; MM Ali Road; Pavamani Road; MCC Cross Road; Rajaji Road; Jail Road; Puthiyangadi-Kakkodi Road; and Francis Road are constitute major roads in the city. Mini Bypass between Meenchanda and West Hill through Mankavu provides the much-needed relief to urban section of NH 17 to a certain extent. The NH 17 Bypass will provide the second level ring to the city. These rings will require adequate radial roads emanating from the concentrated western part of the city, so as to reduce the traffic problems in the narrow central city roads.

The city has a well-developed transport infrastructure. A large number of buses, predominantly run by individual owners and government furnish on the major routes within the city and to nearby locations. Both Kerala State Road Transport Corporation (KSRTC) and private buses runs regular services to many destinations in the state and to the adjacent states from the three bus stands present in the city. Private buses to the suburban and nearby towns ply from the Palayam Bus Stand. Private buses to adjoining districts start from the Mofussil Bus Stand. Buses operated by the KSRTC drive from its bus stand on Mavoor road.

National highway 17 runs along west coast of india connects Kozhikode to Mumbai via Mangalore, Udupi and Goa to the north, and to Kochi in the south. It also connects the city with the other important towns of coastal stretch. National Highway 212 connects Kozhikode city with Kalpetta, Bangalore and Mysore, through the suburbs. SH 29 pass through the city which connects NH 212.

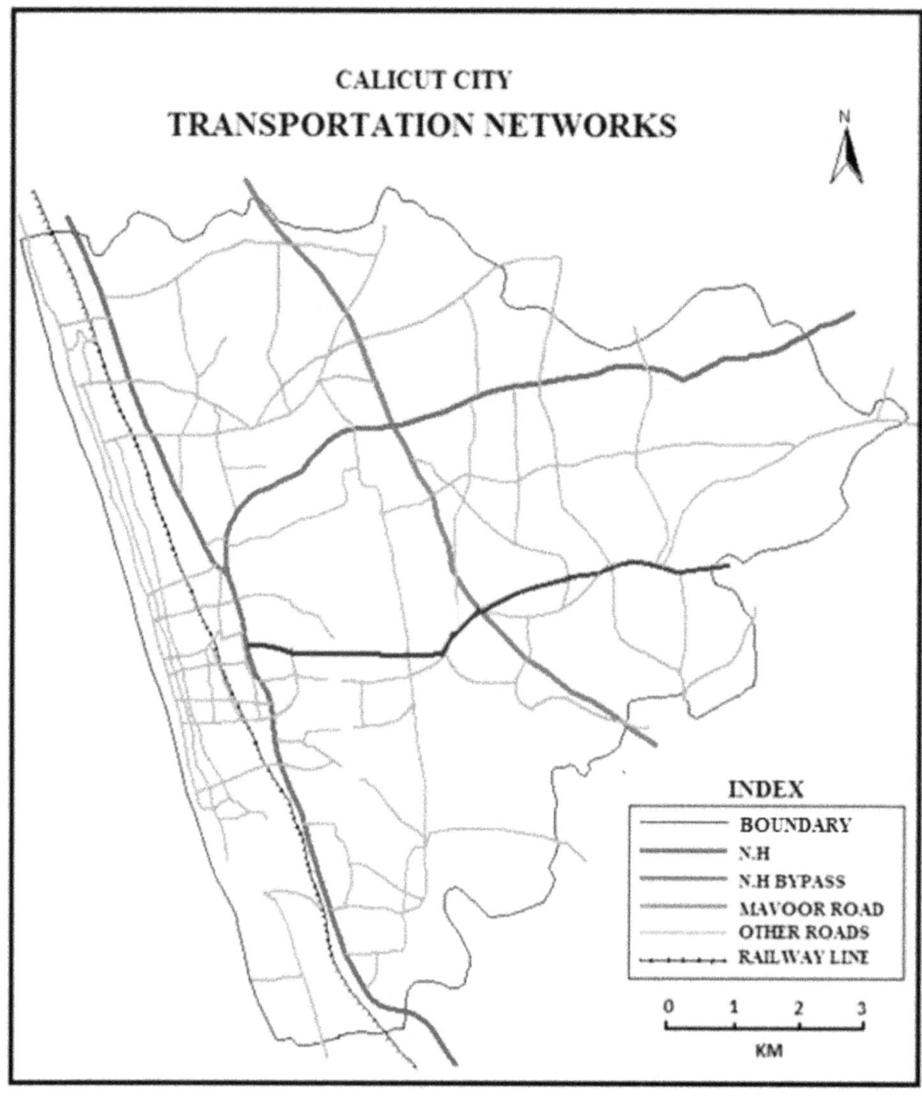

Figure: 4.3 Calicut City Transportation Networks

Kozhikode is presently connected by rail to important cities like the capital city of Kerala ie. Thiruvananthapuram and most of the south Indian cities like Banglooru, Chennai, Coimbatoor, Erode etc. Calicut International Airport which operates regular and frequent services of domestic and international flights is located 22 km from the city at Karipur. Bepur port is also in the vicinity of the city and serves for transportation of bulky goods like timber and furniture.

4.5.2 Education

As a high literacy rate city Calicut have many educational institutions in both government and private sectors. Basic education facility in city is quite well which make the base for higher education. Calicut is home to two premier educational institutions of national importance- the Indian Institutes of Management (IIM), and the National Institute of Technology (NIT). There are many other major institutes in Calicut such as the University of Calicut, Calicut Medical College, College of Nursing, Government Engineering College, Malabar Christian College, Zamorin's Guruvayurappan College, St. Joseph's College Devagiri, Government Arts and Science College, Providence Women's College, Government Homeopathic Medical College, Government Law College, Government College of Teacher Education and Kerala School of Mathematics.

Some research institutes of national importance are also located in or around the city. These include the Indian Institute of Spices Research (IISR), the Centre for Water Resources Development and Management (CWRDM), Western Ghats Field Research Station (Zoological Survey of India) and the Regional Filaria Training and Research Centre.

Table: 4.4 Basic Education Facilities in Calicut City

Type	*Govt.*	*Aided*	*Private*
L.P	22	29	3
U.P	14	19	6
HS/HSE/VHSE	19	14	9

Source: District level database, Kozhikode-2006

4.5.3 Health Care

Health care system in whole Kerala is well advanced in comparison to the other states of India. In the case of Calicut city it is the main medical hub serving whole the Malabar region with one Medical College, Homeopathic college and three other hospitals in government sector and about 30 multi speciality hospitals in private sector like Malabar Institute of Medical Science (MIMS), Baby Memorial hospital, Iqra hospital, National hospital, PVS hospital, Chest hospital, Ashoka hospital and Fathima hospital.

Table: 4.5 Health Care Institutions in Government Sector.

Types	*No. institutions*	*No. of doctors*	*No. beds*
Allopathic	5	84	1682
Homeopathic	1	46	110
Ayurvedic	1	6	110

Source: District level database, Kozhikode-2006

4.5.4 Drainage

Calicut city has an undulated topography which increases from east to west with a ground level variation of about 14 meters. Its natural slope is from east to west with small hilly terrain located in the eastern and central part of the city. Average annual rainfall is about 3,000 mm with an average 115 rainy days a year. A number of natural drainage channels exist in the city. These channels mostly act as secondary drainage outlets and carry both storm as well as wastewater either to the Conolly Canal or to the Arabian Sea. The Conolly Canal is a man made canal connecting the Elathur River in the north and Kallai River in the south of the city and is the main recipient of the surface water runoff. The existing secondary drains are often inadequate to carry the entire storm water runoff of the city particularly during the monsoon season.

Table: 4.6 Drainage Categorization in Calicut City (2006).

Description	*Length (km)*
Primary Drains	95.40
Secondary Drains	102.30
Roadside Internal Drains	52.25
Total length of drains/canals	249.95

Source: Data obtained from the Kozhikode Municipal Corporation.

The Connolly Canal built in 1848 exists without much distortion, running across the heart of the city with a stretch of 11kms. in north south direction. The width of the canal ranges from 6m to 20m in different stretches. The depth of water varies from summer to rainy season, in the peak summer from 0.5m to 2m in heavy rainy season. The canal functions as a drain to reduce flood in the city during the rainy season and also as a navigation channel.

Several areas of the city experience water logging especially in the monsoon season during the period of rainfall of high intensity for extended duration. Some of the major water logged areas are Bus Stand and its adjoining area, Kommeri, Payyanakkal, Thadambattuhazham Vellayil , Puthiyakadavu, Ayyappankavu and Jafar Khan Colony. Movement of traffic and even pedestrian in the roads of these areas are often affected due to water logging.

4.5.5 Sanitation

Existing sanitation facilities in Kozhikode mainly consist of individual septic tanks generally in middle and high-income residential areas, and shallow pit latrines generally in areas with low-income groups. Sullage from houses is discharged into roadside drains and the Conolly Canal, which crosses the central city from north to south. In addition, there are a number of drains that carry sullage from hotels and commercial establishments and directly discharge into the Arabian Sea. Sanitation facilities are not adequate in slum areas. There are 79 slums in the city identified by the department of Town & Country Planning. A detailed account of sanitation condition is included in coming chapter.

4.6 Demographic Aspects

The population over space reflects the physical condition of an area, its history of development and the effectiveness with which the occupant have been able to utilize their cultural tradition. Kerala state is considered as one of the demographically advance state in India, and Calicut city is not an exception from this. The specific characteristics of population density, literacy rate, population growth, sex ratio and the occupational structure in Calicut city are explained below.

Table: 4.7 Ward-wise Population Statistics of Calicut City (2001)

Ward	*Households*	*Population*			*S.C pop.*	*S.T pop.*	*Literacy rate*	*Sex Ratio*
		Male	*Female*	*Total*				
1	1603	5086	5325	10411	281	5	91.89	1047
2	1480	3554	3882	7436	647	0	96.21	1092
3	1524	3490	3911	7401	403	99	96.9	1121
4	1628	4117	4265	8382	489	5	95.4	1036
5	1494	3810	3885	7695	442	17	96.2	1020
6	1524	3527	3802	7329	381	0	96.1	1078
7	1718	3966	4287	8253	646	0	95.5	1081
8	1818	4229	4644	8873	628	2	96.3	1098
9	1565	3418	3656	7074	546	6	96.2	1070
10	1659	3625	3954	7579	533	7	96.8	1091
11	1551	3486	3732	7218	584	9	94.3	1071
12	1630	3709	4118	7827	1099	30	96.2	1110
13	1599	4148	4578	8726	779	1	94.5	1104
14	1545	3872	4103	7975	645	0	93.3	1060
15	1502	3548	3727	7275	407	0	95.6	1050
16	1734	4120	4155	8275	489	4	95.2	1008
17	1817	4522	4480	9002	692	9	94.7	991
18	1562	3720	3979	7699	392	4	97.8	1070
19	1342	3182	3348	6530	329	0	97.6	1052
20	1701	4547	4707	9254	367	2	94.7	1035
21	1680	4382	4668	9050	354	0	95.4	1065
22	1687	4513	4795	9308	212	0	94.1	1062
23	1440	4296	4521	8817	117	0	95.4	1052
24	1513	4023	4178	8201	86	0	95.7	1039
25	1410	4130	4313	8443	96	0	95.2	1044
26	1383	3596	3809	7405	131	0	95.7	1059
27	1736	5160	5529	10689	77	0	94.1	1072
28	1430	4542	4657	9199	106	0	92.6	1025
29	1888	5926	6269	12195	205	4	91.4	1058

Contd......

Table 4.7 Contd....

Ward	Households	Population			S.C pop.	S.T pop.	Literacy rate	Sex Ratio
		Male	Female	Total				
30	1642	5388	5825	11213	139	0	91.2	1081
31	2002	7738	8167	15905	7	0	84.5	1055
32	1837	6526	6755	13281	293	4	91.5	1035
33	1405	3441	3670	7111	172	8	94.8	1067
34	1494	4044	4331	8375	532	4	95.3	1071
35	1556	3932	3996	7928	333	0	95.6	1016
36	1263	3437	3423	6860	554	9	91.4	996
37	1743	4488	4575	9063	840	8	94.7	1019
38	1593	4748	4929	9677	297	10	93.8	1038
39	1665	4511	4961	9472	235	10	94.7	1100
40	1454	3565	3733	7298	388	11	97.5	1047
41	1387	3129	3294	6423	351	5	96.2	1053
42	1552	3778	3801	7579	401	21	95.7	1006
43	1629	3829	4263	8092	383	0	96.8	1113
44	1555	3671	4068	7739	794	6	96.5	1108
45	1322	2971	3188	6159	155	0	97.9	1073
46	1206	2724	3035	5759	434	6	97.4	1114
47	1872	5932	6448	12380	203	9	90.8	1087
48	1599	4925	5332	10257	192	5	90.8	1083
49	1434	3141	3651	6792	125	4	96.8	1162
50	1809	4147	4264	8411	394	18	97.0	1028
51	1346	3579	3691	7270	315	4	92.1	1031
Total	**80528**	**211888**	**224668**	**436556**	**19700**	**346**	**94.3**	**1060**

Source: Department of economics and statistics, Kerala, 2006

4.6.1 Population Growth

As of 2001 India census, Calicut had a population of 436,530 and is the third largest Urban agglomeration in Kerala. Population growth rate shows a continuous increase from 1911 to 1941 and then it began to decline gradually from 1951 onwards. An exception can be noted in 1971, it is not by the rapid change in growth rate but the city population increased significantly due to the extension of the corporation boundary. In recent years the population growth rate is declined from about 18 in 1981 to -1.02 in 2011 (according to the provisional census report) through 6.36 in 1991 and 4.06 in 2001. The population growth rate in the city is more or less following that of the Kerala state.

Table: 4.8 Decadal Population Growth Rate in Calicut City

Years	Population Size			Growth Rate
	Total	*Male*	*Female*	
1901	76981	39986	36995	——
1911	78417	40680	37737	1.87
1921	82334	42527	39807	4.99
1931	99273	51030	48243	20.57
1941	126352	63998	62354	27.28
1951	158724	80069	78655	25.62
1961	192521	97911	94610	21.29
1971	333979	168009	165970	73.47
1981	394447	196628	197819	18.11
1991	419531	206914	212617	6.36
2001	436556	211888	224668	4.06
2011*	432,097	206,494	225,603	-1.02

Source: Census data from 1901-2011
* Provisional

4.6.2 Population Density

The density of population has been calculated in order to find out the spatial variation in the settlement of population within the district and city. The average density of population in the district in 2001 was about 1129 persons/sq.km. It was 1118 persons/sq.km in 1991. The population density of Calicut city, as per the 1981 censes was 4770 and that in 1991 and 2001 are 5077 and 5279 respectively. The average household size of the city as per the 1991 census was 6.2 persons and that of the Kozhikode urban area and Kozhikode district was 6.14 and 5.73 respectively. Density is very high in the older residential areas and is comparatively less in the new residential and peripheral region

4.6.3 Literacy Rate

Calicut district has an average literacy rate of 92.24% (National Avg.: 59.5%) wherein male literacy rate is 96.6% and female literacy rate is 90.6%.and 11% of the total city population is under 6 years of age. In the case of city it is 94.3%. There are adequate facility for the primary and secondary education in both government and private sectors and the people of Calicut are enough bothered about the education of their children. Detail account of literacy can get from the table 4.7

4.6.4 Sex Ratio

Kerala is the only state in India having sex ratio above 0.99. The sex ratio of the district is 1,055 females to every 1000 males, which means there are number women in Kozhikode district than men. It has an average sex ratio of 1,055 females for every 1000 male. Males constitute 49% of the population and females

51%. In the case of city it 1060 females for 1000 males (table: 4.7). in all the wards of the city females outnumber the males except Kovoor and Puthiyara. It is highest in the ward west hill with a ratio of 1162 females for 1000 males.

4.6.5 Occupational Structure

Kozhikode district is not a major industrial area and hence most people are engaged other occupational sectors of economy. According to 2001 census the main workers constitute about 22% of working force of the district. Much young population temporarily migrated to overseas especially to Middle East for employment and the earnings from there is the backbone of the income of many middle class families of the city. According to 2001 census the percentage of working population in the city are only 30.6 comprising 26.8 % main workers and 3.8 % marginal workers. The main workers again distributed as 0.06% of total population is cultivators, 0.11% is agricultural labours, 0.45% engaged in household industry and 26.2% are in other workers category. 69.5% people are non workers. Female work participation rate is comparatively very less which reflects the traditional middle class Indian family system in which females are restricted to go outside for employment, especially in Muslim community.

Table: 4.9 Category-wise Work Participation in Calicut City

Gender	*Population*	*Main Workers*				*Marginal*	*Non*
		Cultivators	*Agri. labours*	*Household Industry*	*Other workers*	*Marginal*	*workers*
Male	211888	226	432	1662	96581	12845	100142
Female	224668	55	28	324	17844	3781	202636
Total	436556	281	460	1986	114425	16626	302778

Source: District census handbook, Kozhikode, 2001

4.7 The People

Calicut has been a multi-ethnic and multi-religious town since the early medieval period. Hindus form the largest religious group, followed by Muslims and Christians. The *Nairs* of Hindu formed the rulers, warriors and landed gentry of Calicut. The *Nairs* also formed the members of the suicide squad (*Chaver*). The aristrocratic *Nairs* had their *Taravad* houses in and around the capital. Several *Nairs* in the city were traders too. The *thiyyas* of Calicut includes *vaidyars*, local militias and traders. There are many aristocratic *Thiyya* families like *Kallingal madom*. The Tamil Brahmins (*Pattar*) are primarily settled around the *Tali* Siva temple. The Gujarati community is settled mostly around the Jain temple in and around the Big Bazaar (Valiyangadi). They owned a large number of establishments, especially textile and sweet shops.

The Muslims of Calicut are known as *Mappilas*, Many of the Muslims living in the historic part of the city follow matrilineal and are noted for their piety. Though Christianity is believed to have been introduced in Kerala in 52 B C, the size of community began to rise only after the arrival of the Portuguese towards the close of the 15th century. The *Mappila* community of Calicut acted as an important support base for the city's military, economic and political affairs. They were settled primarily in Kuttichira and Idiyangara. Their aristocratic dwelling houses were similar to the *tharavad* houses of the *Nairs* and the *Thiyyas.*

Pre-modern Calicut was already teeming with people of several communities and regional groups. Most of these communities continued to follow their traditional occupations and customs till the 20th century. These included *Kosavan* (potter), *Mannan* or *Vannan* (washerman), *Pulayan* (agricultural worker), *Chaliyan* (weaver), *Chetti* (merchant), *Ganaka* (astrologer), *Vettuvan* (salt-maker), *Paanan* (sorcerer). A number of Brahmins too lived in the city mostly around the Hindu temples. The story of Calicut is the story of all the communities. However, the social groups that exerted the greatest influence in the history of Calicut have been the *Nairs, Thiyyas* and the *Mappilas.*

4.7.1 Social Organization

Neighborhood Groups (NHG), Area Development Societies (ADS) and Community Development Societies are contributing three tier Community Based Organizations in Calicut city. *Kudumbashree* is the back bone of this emerging strong social capital structure having crucial functions on the design and development of different levels of programs for the poor and socially and economically backward communities. There are 44 ADS and 700 NHG's in the city. There are many resident welfare associations in Kozhikode which focus on household problems which operate by the residents itself in a democratic way. There is a Federation of Residents Association (FRAT) representing these associations. There are also consumer groups, citizens groups, merchants and industrialists associations and many non political and political community level organizations for the welfare of the members and the family of the members which create a civic conscious for the social and cultural development of city.

4.7.2 Malabar Mahotsavam

Malabar Mahotsavam is an art and cultural festival that aims to recapture and revive the glory of the tradition and heritage once prevalent in the Malabar regions. The event is held at different venues like the beach and the Mananchira grounds in the heart of Calicut town, in the month of January every year. It is a common platform of artists from all over the state and a media to cultural integration. The Festival includes Classical and folk arts, agricultural and handloom/handicrafts exhibitions cum sales, shopping festivals etc. This mega fest encourages the local artists and gives the local population an exposure to the varied art forms of India, and exploits the immense tourism potential of the region. It is also aimed to create an atmosphere of communal harmony, as Malabar is known for the peaceful co-existence of many religious faiths.

4.8 The Economy

Calicut is one of the main commercial cities of Kerala and its economy is mainly business oriented. With good connectivity through road, rail water and air it is currently the major trade hub of North Kerala. Calicut occupies a nodal position in the Malabar Coast and has an undisputed economic standing of its own being the trade centre for a large and resourceful hinterland. It is the market place of hill products like pepper, cardamom, etc. It also has large timber yards along the banks of the Kallayi River. Large number of malls and business establishment are mushroomed in recent years. A considerable proportion of the middle age male population are employed in the Middle Eastern countries, and their remittances to home are an important part of the local economy.

4.8.1 Shipbuilding Industry

From ancient times, the small coastal town of Beypore near Calicut city which is an important harbour in the Malabar region has been synonymous with the traditional ship building culture of Kerala. The art and science of *uru* making learnt from the Arab traders thousands of years back when they landed at the port for sea trade. These traditional Arabian trading vessels were called 'dhows'. The builders of this traditional country craft are referred to as *'khalasis'*. These adept ship builders had deep knowledge about their craft. The *urus* still have a swelling market and continues to draw buyers from across the Arabian Sea. The availability of good timber and the skilled craftsmen in ship building led the Arab merchants to place orders for constructing dhows to the craftsmen of Malabar in north Kerala. Thus the foundation for the ship-building industry was laid, and the tradition still continues, although on a smaller scale.

4.8.2 Agriculture

Calicut city and surrounding areas have a rich heritage in agriculture as it was a port city famous for pepper & spices trade for a long period. Agriculture plays a major role in the district's economy. The crops like coconut, paddy, banana, tubers and other spices and tree crops are cultivated in the region. The other important crops are ginger, turmeric, pepper, nuts, cashew, rubber, cocoa, arecanut and tapioca.

4.8.3 Fisheries

Calicut is endowed with a coast line which offers enormous resources for development of fisheries. The region is rich in brackish water area and there is great scope for shrimp farming too. In the coastal belt, fishing is the main occupation of a large number of people. Large numbers of fishermen are directly involved in fishing activities. The allied industries such as ice plant, freezing and processing units also provide employment to the people. As a whole, the fishing industry makes a sizeable contribution to the economy of the city, district as well as the State.

5
GENERAL ENVIRONMENT IN CALICUT CITY

The city of Calicut with its medium population, coastal location, and unique cultural and social aspects have very much advanced general environment in comparison with the same class Indian cities. In this chapter the author tried to outline the general environment in the city in terms of domestic water, sanitation, housing structure etc. The entire data and description is based on the primary source obtained by field survey conducted by the author during February to September 2010 with the help of prescribed schedule.

Environmental problems and the challengers is the unwanted guest with the urbanization. The older residential areas are the most congested and unhygienic region in Indian cities in general and Calicut city in particular, due to several reasons. Calicut city also confronts a number of urban environmental challenges along with the existing common urban environmental threats by the geographical features, rapid changing of life style and unplanned urbanization. Calicut city has been expanded and developed by filling low lying areas. A considerable part of the built up areas in city has been constructed by filling the wetlands and marshy lands and the land filling process is going on at an accelerated rate.

The human settlement of Calicut is started from the seashores and nearby places. The first external who managed a settlement cluster in Calicut city may be the migrants from the Kutch region of Gujarat came here for trade, they settled near seashores along with their business establishments. The Arabs, English, Portuguese and the Dutch also made their settlements near the seashore but most them are not sustaining in their original forms. The coastal areas are became densely populated and the population stress led to the stagnation of sewerages and out flow of the same especially during rainy season which made the area unhygienic and polluted the drinking water, air and soil. The congestion with high housing density of the city led to the contamination of the wells- the prime source of domestic water- as one's well is situated very near to the sanitary soak-pits of neighbouring houses or of the same house due to small holding size. Domestic wastes, wastes from shops, hotels, and hospitals are also put in to the sewage which constitutes organic wastes as the chief component and they undergo bacterial decomposition in the water thereby generating many toxic gases and also generate medium to the easy growth of pathogens. The small scale industries like coir factory, paint factory, soap factory and oil mills discharge their crude effluents into sewages as most of them do not have waste treatment systems. Most of the sewages are opening to the Lakshadweep Sea either directly or through Canoli canal and Kallai river which pollutes the seawater also.

5.1 Source of Drinking Water

The quality of drinking-water is a powerful environmental determinant of health. People are increasingly concerned about the safety of their drinking water. The health effects of some contaminants in drinking water are not well understood. Assurance of drinking-water safety is a foundation for the prevention and control of waterborne diseases. The most affected are the populations in developing countries, living in extreme conditions of poverty, normally peri-urban dwellers or rural inhabitants. Among the main problems which are responsible for this situation are: lack of priority given to the sector, lack of financial resources, lack of sustainability of water supply and sanitation services, poor hygiene behaviours, and inadequate sanitation in public places.

Private water sources, including wells, are also not regulated by drinking water standards, and the owner must take steps to test and treat the water as needed to avoid possible health risks. There is another major problem in the core city regions with the small land holdings i.e. one's well may be located very near to the soak pits or septic tanks of the neighbouring house, of the same house which will lead the pollution of the well water.

Table: 5.1 Household Drinking Water Facilities in Calicut in Percentages.

Ward No.	*Ward Name*	*Private*	*Corp. Connection*	*Public*
1	New Bazar	48.0	40.0	12.0
2	Edakkat	43.5	56.5	0.0
3	Easthill	72.0	28.0	0.0
4	Kuruvisseri	72.0	28.0	0.0

Contd....

Table 5.1 Contd...

Ward No.	Ward Name	Private	Corp. Connection	Public
5	Malaparamba	59.1	40.9	0.0
6	Vengeri	69.6	30.4	0.0
7	Kannadikkal	57.7	38.5	3.8
8	Paroppadi	74.1	25.9	0.0
9	Chevarambalam	78.3	21.7	0.0
10	Kudilthodu	76.0	24.0	0.0
11	Cheveyur	73.9	26.1	0.0
12	Silverhills	72.0	28.0	0.0
13	Poolakkadavu	75.0	25.0	0.0
14	Moozhikkal	73.9	26.1	0.0
15	Chelavoor	69.6	30.4	0.0
16	Mayanadu	47.8	47.8	4.3
17	Kovoor	63.0	33.3	3.7
18	Nellikkode	73.9	26.1	0.0
19	Potamal	65.0	35.0	0.0
20	Kommeri	69.2	30.8	0.0
21	Pokkunnu	60.0	40.0	0.0
22	Mankavu	76.0	24.0	0.0
23	Kinasseri	72.7	27.3	0.0
24	Thiruvannur	73.9	26.1	0.0
25	Kallai	63.6	36.4	0.0
26	Panniyankara	81.0	19.0	0.0
27	Meenchanda	73.1	26.9	0.0
28	Koya Valappu	54.5	27.3	18.2
29	Payyanakkal	46.4	32.1	21.4
30	Chakkumkadavu	44.0	24.0	32.0
31	Pallikkandi	36.7	36.7	26.7
32	Idiangara	50.0	35.7	14.3
33	Chalappuram	45.5	54.5	0.0
34	Azhchavattam	59.1	40.9	0.0
35	Kuthiravattam	34.8	65.2	0.0
36	Puthiyara	63.2	36.8	0.0
37	Palayam	53.8	42.3	3.8
38	Big Bazar	66.7	29.2	4.2
39	Vellayil South	28.0	56.0	16.0
40	Thiruthiyad	59.1	40.9	0.0
41	Kotooli South	66.7	33.3	0.0

Contd....

Table 5.1 Contd...

Ward No.	Ward Name	Private	Corp. Connection	Public
42	Kotooli North	52.2	47.8	0.0
43	Civil Station	66.7	33.3	0.0
44	Eranchippalam	65.2	34.8	0.0
45	Karaparamba	65.0	35.0	0.0
46	Nadakkavu	61.1	38.9	0.0
47	Vellayil North	32.1	50.0	17.9
48	Thoppayil	33.3	50.0	16.7
49	Chakkorath Kulam	63.6	36.4	0.0
50	Westhill	77.8	22.2	0.0
51	Varaykal	55.0	45.0	0.0
	Averages	**61.1**	**35.1**	**3.8**

Source: Field Survey, 2010

As a low laying coastal city intensively drained by rivers, canals and small fresh water gullies, Calicut city have basically no major problem of portable water. But some coastal areas are facing the problem of saline water, and the inner city region facing the problem of pollution of surface and sub surface water. Majority of households using own open and closed wells for their domestic water. Some elevated land facing the scarcity of subsurface water while some other low lying areas are facing the problem of salinity and hence they using the corporation connection. Most of the high class population and middle class population are using both the private well and corporation connection. Even in some other areas people use public taps provided by the corporation for their needs and other public sources like open well ponds and fresh water gullies etc.

Out of 1211 households surveyed 736 (61.1 %) households are using their own open or closed well as their primary source, 424 (35.1%) households are using corporation connection as prime source and 51 households (3.8 %) using public sources. More than seventy per cent of the households of the wards Easthill, Kuruvisseri, Vengeri, Paropadi, Cheverambalam, Kudilthode, Chevayoor, Silver hills, Poolakkadavu, moozhikkzl, Nellikkode, Mankavu, kinasseri, Thiruvannur, Panniyankara, Meenchanda and West hill are using private open wells as prime source of drinking and domestic water. Coastal wards with congested residential areas are mostly depends corporation connection or the public taps or other public sources. Koya valappu, Payyanakkal, Chakkumkadavu, Pallikkandi, Idiangara, Vellayil North, Vellayil South and Thoppayil are most vulnerable wards in this respect.

5.2 Sanitation

Sanitation is very important for modern man especially in urban places. A large proportion of the developing world's population lack improved sanitation facilities and millions of people still use unsafe drinking water sources. Inadequate access to safe water and sanitation services, coupled with poor hygiene practices, kills and sickens thousands of children every day. Lack of private and decent sanitation facilities largely affect the health condition of the ladies than men, and it will directly affect their children which transmit through the generation. So sanitation is very important for a well and decent society.

Table: 5.2 Household Level Sanitation Facilities in Calicut in Percentages.

Ward No.	Ward Name	Inside	Outside	Total	Open land
1	New Bazaar	72.0	28.0	100.0	0.0
2	Edakkat	73.9	26.1	100.0	0.0
3	Easthill	80.0	20.0	100.0	0.0
4	Kuruvisseri	84.0	16.0	100.0	0.0
5	Malaparamba	86.4	13.6	100.0	0.0
6	Vengeri	82.6	17.4	100.0	0.0
7	Kannadikkal	65.4	34.6	100.0	0.0
8	Paroppadi	85.2	14.8	100.0	0.0
9	Chevarambalam	82.6	17.4	100.0	0.0
10	Kudilthodu	84.0	16.0	100.0	0.0
11	Cheveyur	91.3	8.7	100.0	0.0
12	Silverhills	88.0	12.0	100.0	0.0
13	Poolakkadavu	75.0	25.0	100.0	0.0
14	Moozhikkal	73.9	26.1	100.0	0.0
15	Chelavoor	69.6	30.4	100.0	0.0
16	Mayanadu	78.3	21.7	100.0	0.0
17	Kovoor	66.7	33.3	100.0	0.0
18	Nellikkode	73.9	26.1	100.0	0.0
19	Potamal	70.0	30.0	100.0	0.0
20	Kommeri	61.5	38.5	100.0	0.0
21	Pokkunnu	52.0	48.0	100.0	0.0
22	Mankavu	72.0	28.0	100.0	0.0
23	Kinasseri	72.7	27.3	100.0	0.0
24	Thiruvannur	82.6	17.4	100.0	0.0
25	Kallai	81.8	18.2	100.0	0.0
26	Panniyankara	85.7	14.3	100.0	0.0
27	Meenchanda	80.8	19.2	100.0	0.0
28	Koya Valappu	31.8	54.5	86.4	13.6
29	Payyanakkal	32.1	57.1	89.3	10.7
30	Chakkumkadavu	16.0	72.0	88.0	12.0
31	Pallikkandi	23.3	63.3	86.7	13.3
32	Idiangara	50.0	42.9	92.9	7.1

Contd....

Table: 5.2 Contd...

Ward No.	Ward Name	Inside	Outside	Total	Open land
33	Chalappuram	63.6	36.4	100.0	0.0
34	Azhchavattam	72.7	27.3	100.0	0.0
35	Kuthiravattam	82.6	17.4	100.0	0.0
36	Puthiyara	78.9	21.1	100.0	0.0
37	Palayam	50.0	50.0	100.0	0.0
38	Big Bazar	54.2	45.8	100.0	0.0
39	Vellayil South	36.0	56.0	92.0	8.0
40	Thiruthiyad	72.7	27.3	100.0	0.0
41	Kotooli South	71.4	28.6	100.0	0.0
42	Kotooli North	69.6	30.4	100.0	0.0
43	Civil Station	91.7	8.3	100.0	0.0
44	Eranchippalam	82.6	17.4	100.0	0.0
45	Karaparamba	75.0	25.0	100.0	0.0
46	Nadakkavu	88.9	11.1	100.0	0.0
47	Vellayil North	42.9	46.4	89.3	10.7
48	Thoppayil	37.5	54.2	91.7	8.3
49	Chakkorath Kulam	72.7	27.3	100.0	0.0
50	Westhill	85.2	14.8	100.0	0.0
51	Varaykal	80.0	20.0	100.0	0.0
	Averages	**69.3**	**29.1**	**98.4**	**1.6**

Source: Field Survey, 2010

In Calicut city most of the households have their own sanitation facility. A majority of households have sanitation facility inside their houses. This facility is mainly available in the RCC houses of professionals and business classes of high and medium socio economic condition. Majority of RBC houses have also their own sanitation facilities but outside the houses. Sanitation facility is of septic tanks and soaking pit type. Moreover a minute proportion of households have only common sanitation facility either provided by government or constructed collectively by a small group of house. Another minority of population, especially in coastal areas, have no access to established sanitation facilities and they using open lands for their sanitary needs.

From the total households 829 (69.3 per cent) households have sanitation facility inside their dwellings. 360 (29.1 per cent) have the facility outside the houses and only 22(1.6 per cent) have no accessible sanitation facility their own. Most of the wards have hundred per cent sanitation facility except Koya valppu (86.4%), Pallikkandi (86.7%), Chakkum kadavu (88%), Payyanakkal (89.3%), Vellayil north (89.3%), Thoppayil (91.7%), Vellayil South (92%), and Idiangara (92.9%).

5.3 Nature of Family

A family is a group of people affiliated by affinity or co-residence. In most societies it is the principal institution for the socialization of children. The different types of families occur in a wide variety of settings, and their specific functions and meanings depend largely on their relationship to other social institutions. Nuclear family consists of husband, wife and their children. The concept of joint family where all the family members like aunt, uncle, cousins and grandparents live together contradicts with the concept of nuclear family

Nature of family has well defined effect in the formation of well society. Both the nuclear and joint families have their own advantages and shortcomings. So analysis which related to the social environment must consider these two systems of family organization. In Calicut city the traditional families of Muslims and caste Hindus are following the joined family systems and are prominent in the older settlements. The progressive communities and professionals are practicing nuclear family system and are prominent in relatively new residential areas.

Table: 5.3 Type of Family in Calicut City in Percentage

Ward No.	Wards	Nuclear	Joint
1	New Bazar	64.0	36.0
2	Edakkat	91.3	8.7
3	Easthill	88.0	12.0
4	Kuruvisseri	76.0	24.0
5	Malaparamba	81.8	18.2
6	Vengeri	82.6	17.4
7	Kannadikkal	88.5	11.5
8	Paroppadi	77.8	22.2
9	Chevarambalam	91.3	8.7
10	Kudilthodu	84.0	16.0
11	Cheveyur	95.7	4.3
12	Silverhills	80.0	20.0
13	Poolakkadavu	75.0	25.0
14	Moozhikkal	82.6	17.4
15	Chelavoor	82.6	17.4
16	Mayanadu	82.6	17.4
17	Kovoor	74.1	25.9
18	Nellikkode	91.3	8.7
19	Potamal	70.0	30.0
20	Kommeri	76.9	23.1
21	Pokkunnu	76.0	24.0
22	Mankavu	84.0	16.0
23	Kinasseri	72.7	27.3

Contd....

Table : 5.3 Contd....

Ward No.	Wards	Nuclear	Joint
24	Thiruvannur	87.0	13.0
25	Kallai	72.7	27.3
26	Panniyankara	71.4	28.6
27	Meenchanda	80.8	19.2
28	Koya Valappu	72.7	27.3
29	Payyanakkal	64.3	35.7
30	Chakkumkadavu	72.0	28.0
31	Pallikkandi	60.0	40.0
32	Idiangara	46.4	53.6
33	Chalappuram	86.4	13.6
34	Azhchavattam	86.4	13.6
35	Kuthiravattam	87.0	13.0
36	Puthiyara	89.5	10.5
37	Palayam	76.9	23.1
38	Big Bazar	62.5	37.5
39	Vellayil South	80.0	20.0
40	Thiruthiyad	77.3	22.7
41	Kotooli South	85.7	14.3
42	Kotooli North	82.6	17.4
43	Civil Station	91.7	8.3
44	Eranchippalam	73.9	26.1
45	Karaparamba	85.0	15.0
46	Nadakkavu	77.8	22.2
47	Vellayil North	75.0	25.0
48	Thoppayil	79.2	20.8
49	Chakkorath Kulam	77.3	22.7
50	Westhill	77.8	22.2
51	Varaykal	80.0	20.0
	Average	**79.0**	**21.0**

Source: Field Survey, 2010

Out of 1211 households surveyed 952 (79 per cent) practicing nuclear family system and the rest 259 (21per cent) practicing joint family system. The wards of Chevayur (95.7%), Edakkat(91.3%), Nellikkode (91.3%), Civil station (91.7%), Puthiyara (89.5%), Kannadikkal (88.5%) and East hill (88%) are leading in the composition of the nuclear family system. And the wards of Velleyil North (25%), New Bazar (36%), Poolakkadavu (25%), Kovoor (25.9%), Potammal (30%), Kinasseri (27.3%), Kallai (27.3%), Panniyankara (28.6%), Koya Valappu (27.3%), Payyanakkal (35.7%), Chakkumkadavu (28%), Pallikkandi (40%), Idiangara (53.6%), Big Bazar

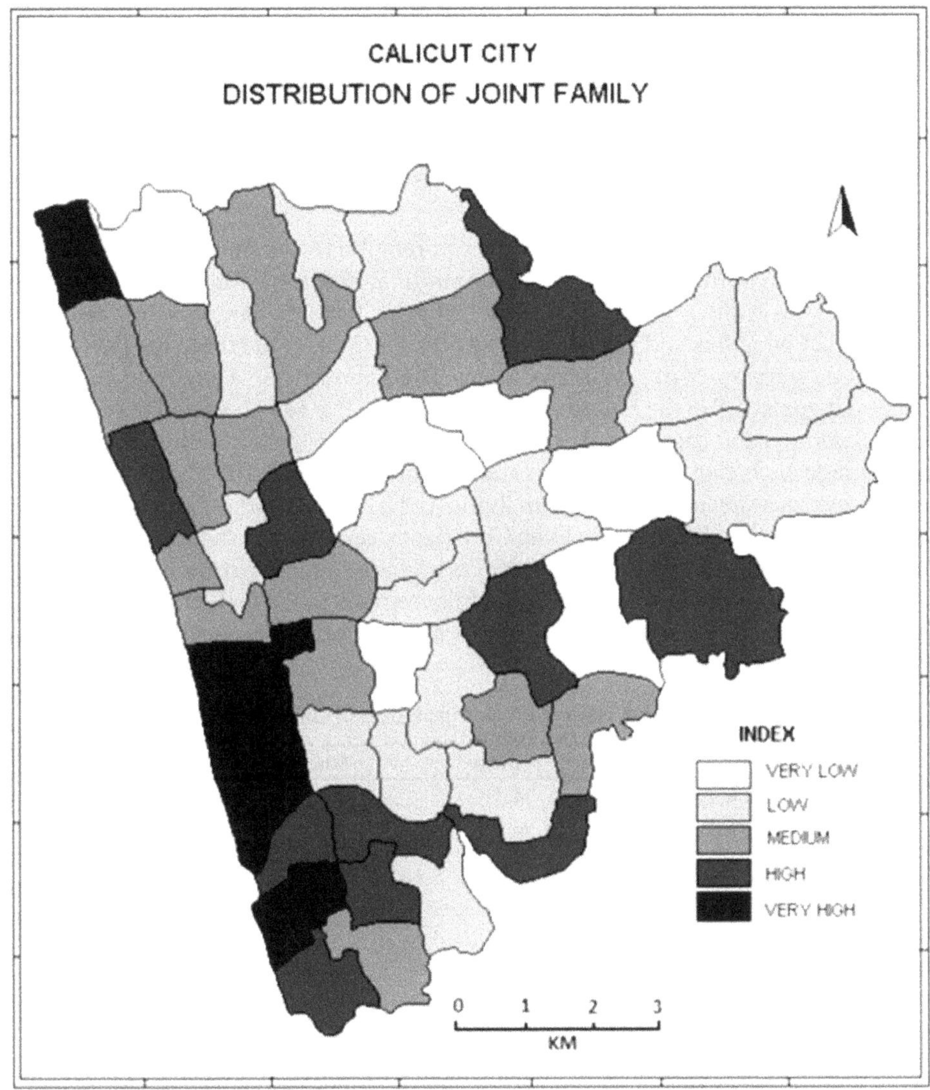

Figure: 5.1 Calicut City Distribution of Joint Family

(37.5%), Eranchippalam (26.1%) are leading in the percentage of the joint families. The areas or wards which showing the nuclear family are those areas which is mostly resided by the Christian and Hindu population. The wards which show the joint family system are mostly resided by the Muslim population as they promoting the joint family system especially the traditional families of older residential areas in the city.

5.4 Religious Composition

Freedom of Religion is one of the most basic rights of human being and is a much debated subject matter. Sometimes the right of one's religious practice arise against those little social rules our society has developed over the time. But we want to take care to keep the social harmony along with all diversity. Social harmony means living with unity, mutual reciprocity and mutual respect, beyond class, caste, creed and gender barriers. Sin or violence emerges whenever this harmony is disturbed. We want to create social harmony among many people who are different in many ways to develop a society and a nation by avoiding some silly things and giving respect to other creed and religion.

Calicut has been a multiethnic and multi-religious town since the early medieval period with an enthusiastic social and communal harmony. Mainly three religions constitute the entire population in which Hindus form the largest religious group, followed by Muslims and Christians. The *Nair* community of Hindu religion formed the rulers, warriors of pre-colonial Calicut. The Zamorin had a ten thousand strong *Nair* bodyguard called the *Kozhikkottu pathinaayiram* (The Ten Thousand of Kozhikode) who defended the capital and supported the administration within the city. The aristocratic *Nairs* had their settled in and around the capital. The Muslims of Calicut are known as *Mappilas*, most of them living in the historic part of the city and follow matrilineal system of succession. Even Christianity came earlier; the size of community began to rise only after the arrival of the Portuguese. A few Christians of Travancore and Cochin have lately migrated to the hilly regions of the Calicut district and are settled there, from there they began to migrate to the cities especially to Calicut for better facilities and employment.

Table: 5.4 Ward-wise Religious Composition of Calicut in Percentages

Ward No.	*Ward Name*	*Hindus*	*Muslims*	*Christians*
1	New Bazar	24.0	72.0	4.0
2	Edakkat	78.3	17.4	4.3
3	Easthill	60.0	20.0	20.0
4	Kuruvisseri	64.0	20.0	16.0
5	Malaparamba	63.6	13.6	22.7
6	Vengeri	69.6	21.7	8.7
7	Kannadikkal	46.2	30.8	23.1
8	Paroppadi	33.3	40.7	25.9
9	Chevarambalam	47.8	26.1	26.1
10	Kudilthodu	56.0	16.0	28.0
11	Cheveyur	52.2	17.4	30.4
12	Silverhills	56.0	20.0	24.0
13	Poolakkadavu	37.5	41.7	20.8
14	Moozhikkal	39.1	56.5	4.3
15	Chelavoor	43.5	47.8	8.7
16	Mayanadu	56.5	39.1	4.3
17	Kovoor	51.9	37.0	11.1
18	Nellikkode	60.9	8.7	30.4
19	Potamal	65.0	35.0	0.0

Contd....

Table: 4.5 Contd....

Ward No.	Ward Name	Hindus	Muslims	Christians
20	Kommeri	57.7	38.5	3.8
21	Pokkunnu	56.0	36.0	8.0
22	Mankavu	80.0	16.0	4.0
23	Kinasseri	40.9	54.5	4.5
24	Thiruvannur	91.3	8.7	0.0
25	Kallai	50.0	50.0	0.0
26	panniyankara	42.9	57.1	0.0
27	Meenchanda	57.7	38.5	3.8
28	Koya Valappu	45.5	54.5	0.0
29	Payyanakkal	46.4	50.0	3.6
30	Chakkumkadavu	36.0	64.0	0.0
31	Pallikkandi	10.0	90.0	0.0
32	Idiangara	3.6	96.4	0.0
33	Chalappuram	77.3	18.2	4.5
34	Azhchavattam	72.7	18.2	9.1
35	Kuthiravattam	65.2	17.4	17.4
36	Puthiyara	63.2	26.3	10.5
37	Palayam	50.0	46.2	3.8
38	Big Bazar	25.0	70.8	4.2
39	Vellayil South	44.0	48.0	8.0
40	Thiruthiyad	59.1	22.7	18.2
41	Kotooli South	81.0	9.5	9.5
42	Kotooli North	78.3	8.7	13.0
43	Civil Station	45.8	25.0	29.2
44	Eranchippalam	47.8	30.4	21.7
45	Karaparamba	55.0	20.0	25.0
46	Nadakkavu	50.0	22.2	27.8
47	Vellayil North	46.4	42.9	10.7
48	Thoppayil	50.0	37.5	12.5
49	Chakkorath Kulam	63.6	13.6	22.7
50	Westhill	74.1	11.1	14.8
51	Varaykal	85.0	10.0	5.0
	Average	**54.1**	**34.0**	**11.9**

Source; Field Survey, 2010

The religious composition of the sampled households shows the Hindus contributing 645 (54.1 per cent) out of 1211, Muslims contributing 423 (34.0 per cent) household and the Christians143 (11.9 per cent). The wards of Tiruvannur, Mankavu, kotooli South, Varaikal, Kotooli North, Edakkt, Chalappurom and Azhchavatom have a lion share of population following Hindu religion. At the same time New Bazaar, Mozhikkal, Kinasseri, Panniyankara, Koya valappu, Chakkum kadavu, Pallikandi, Idiangara and Big Bazaar have majority of Muslim population. The Christian population is mainly concentrated on the Wards Silver hills, Chevayur, Kudilthodu, Nellikkode, Nadakkavu and Civil station.

5.5 Ownership of Houses

A home is a place of residence in which an individual or a family can live and store personal property. Most of the modern households contain sanitary facilities. Human being attached to their home not only in terms of physical location, but it brings a mental or emotional state of comfort. As a social animal, a person's home has been physiologically influence on their behaviour, emotions, and overall mental health. So having a home is very much important in the physical and social environment of any city and village. Ownership of houses or housing security is very important in social wellbeing as having a shelter is one of the important needs of human being. In Calicut city there are some migrated population for the education purpose of their children or for their own government job and high profile jobs in private sector, even for both. Such population owing to their intention to leave the city after their requirements, prefers to settle in the rented houses in elite class residential areas like Silver hills, Malaparamba and Chevayor.

Table: 5.5 Ownership of House in Calicut City in Percentages.

Ward No.	*Wards*	*Owned*	*Rented*
1	New Bazar	96.0	4.0
2	Edakkat	91.3	8.7
3	Easthill	88.0	12.0
4	Kuruvisseri	92.0	8.0
5	Malaparamba	77.3	22.7
6	Vengeri	91.3	8.7
7	Kannadikkal	96.2	3.8
8	Paroppadi	92.6	7.4
9	Chevarambalam	91.3	8.7
10	Kudilthodu	88.0	12.0
11	Cheveyur	82.6	17.4
12	Silverhills	84.0	16.0
13	Poolakkadavu	95.8	4.2
14	Moozhikkal	95.7	4.3
15	Chelavoor	100.0	0.0
16	Mayanadu	95.7	4.3
17	Kovoor	92.6	7.4

Contd...

Table : 5.5 Contd....

Ward No.	Wards	Owned	Rented
18	Nellikkode	91.3	8.7
19	Potamal	90.0	10.0
20	Kommeri	92.3	7.7
21	Pokkunnu	92.0	8.0
22	Mankavu	88.0	12.0
23	Kinasseri	95.5	4.5
24	Thiruvannur	91.3	8.7
25	Kallai	86.4	13.6
26	Panniyankara	90.5	9.5
27	Meenchanda	92.3	7.7
28	Koya Valappu	81.8	18.2
29	Payyanakkal	89.3	10.7
30	Chakkumkadavu	96.0	4.0
31	Pallikkandi	90.0	10.0
32	Idiangara	92.9	7.1
33	Chalappuram	90.9	9.1
34	Azhchavattam	95.5	4.5
35	Kuthiravattam	87.0	13.0
36	Puthiyara	94.7	5.3
37	Palayam	73.1	26.9
38	Big Bazar	75.0	25.0
39	Vellayil South	88.0	12.0
40	Thiruthiyad	95.5	4.5
41	Kotooli South	95.2	4.8
42	Kotooli North	91.3	8.7
43	Civil Station	83.3	16.7
44	Eranchippalam	91.3	8.7
45	Karaparamba	95.0	5.0
46	Nadakkavu	88.9	11.1
47	Vellayil North	89.3	10.7
48	Thoppayil	91.7	8.3
49	Chakkorath Kulam	90.9	9.1
50	Westhill	88.9	11.1
51	Varaykal	95.0	5.0
	Average	**90.4**	**9.6**

Source: Field Survey, 2010.

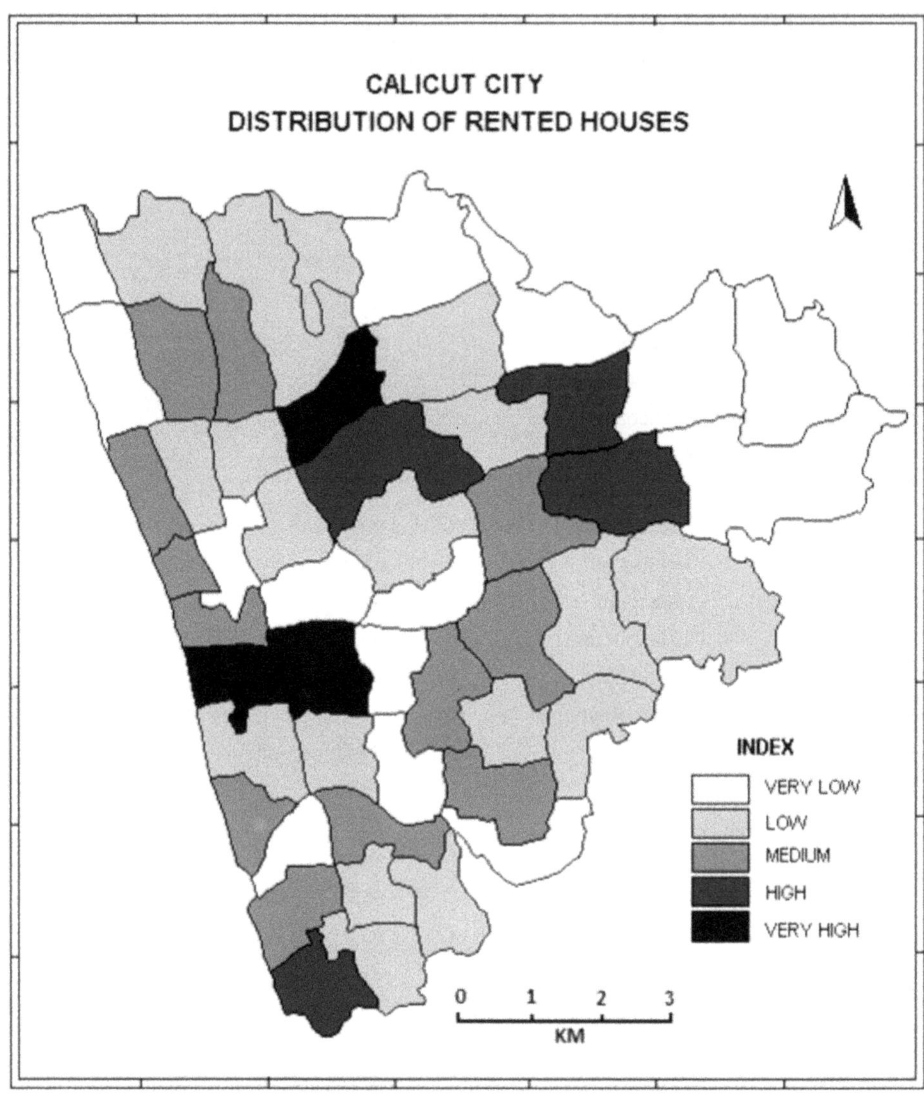

Figure: 5.2 Calicut City Distribution of Rented Houses

Out of 1211 households sampled 1094 houses (90.4 per cent) are owned houses and the rest 117(9.6 percent) is rented houses or cottages. Varaikal, Karaparamba, Kotooli North, Kotooli south, Kuthiravattom, Chakkum kadavu, Thiruvannur, Mayanadu, Chelvur, Mozhikkal, Poolakkadavu, Kannadikkal and New Bazaar have more than 95 percent owned houses. At the same time the percentage of rented houses is comparatively high in Malaparamba (22.7%), Chevayur (17.4%), Silver hills (16%), Koya valappu (18.2%), Palayam (26.9%), Big Bazaar (25%) and Civil Station (16.7%). The proportion of rented houses and owned houses denotes the socio-economic condition and urban environment in the respective wards. It may also due to the presence of prestigious institutions and famous companies in both government and private sector.

5.6 Qualitative Aspects of Houses

A house is a building or structure that has the ability to carry habitation of human beings. The term house includes many kinds of dwellings ranging from rudimentary huts of nomadic tribes to complex structures composed of many houses and facilities. A home will be safe, healthy, and comfortable for the inhabitants in terms of structure, stability, living space and open space. A detached house means that the building does not share an inside wall with any other house or dwelling. It has only outside walls and does not touch any other dwelling. Most single-family homes are built on plots larger than the structure itself, adding an area surrounding the house, which is commonly called as courtyard or garden. Generally Calicut city and Kerala have a culture of detached houses with enough open spaces all around the dwellings, but the congestion and the small land holdings are forced to compromise with this in recent years at least in some areas of the Calicut city.

Table: 5.6 Qualitative Aspects of Houses in Calicut City in Percentages.

Ward No.	*Wards*	*R.C.C*	*R.B.C*	*Huts*
1	New Bazar	64.0	36.0	0.0
2	Edakkat	65.2	34.8	0.0
3	Easthill	84.0	16.0	0.0
4	Kuruvisseri	68.0	32.0	0.0
5	Malaparamba	72.7	27.3	0.0
6	Vengeri	78.3	21.7	0.0
7	Kannadikkal	65.4	34.6	0.0
8	Paroppadi	74.1	25.9	0.0
9	Chevarambalam	87.0	13.0	0.0
10	Kudilthodu	84.0	16.0	0.0
11	Cheveyur	91.3	8.7	0.0
12	Silverhills	88.0	12.0	0.0
13	Poolakkadavu	75.0	25.0	0.0
14	Moozhikkal	73.9	26.1	0.0
15	Chelavoor	69.6	30.4	0.0
16	Mayanadu	78.3	21.7	0.0
17	Kovoor	70.4	29.6	0.0
18	Nellikkode	69.6	30.4	0.0
19	Potamal	75.0	25.0	0.0
20	Kommeri	61.5	38.5	0.0
21	Pokkunnu	56.0	44.0	0.0
22	Mankavu	76.0	24.0	0.0
23	Kinasseri	77.3	22.7	0.0
24	Thiruvannur	65.2	34.8	0.0

Contd...

Table: 5.6 Contd...

Ward No.	Wards	R.C.C	R.B.C	Huts
25	Kallai	81.8	18.2	0.0
26	Panniyankara	85.7	14.3	0.0
27	Meenchanda	80.8	19.2	0.0
28	Koya Valappu	36.4	54.5	9.1
29	Payyanakkal	35.7	57.1	7.1
30	Chakkumkadavu	20.0	72.0	8.0
31	Pallikkandi	30.0	56.7	13.3
32	Idiangara	25.0	67.9	7.1
33	Chalappuram	63.6	36.4	0.0
34	Azhchavattam	72.7	27.3	0.0
35	Kuthiravattam	82.6	17.4	0.0
36	Puthiyara	78.9	21.1	0.0
37	Palayam	53.8	46.2	0.0
38	Big Bazar	33.3	66.7	0.0
39	Vellayil South	40.0	52.0	8.0
40	Thiruthiyad	72.7	27.3	0.0
41	Kotooli South	76.2	23.8	0.0
42	Kotooli North	69.6	30.4	0.0
43	Civil Station	87.5	12.5	0.0
44	Eranchippalam	82.6	17.4	0.0
45	Karaparamba	75.0	25.0	0.0
46	Nadakkavu	88.9	11.1	0.0
47	Vellayil North	42.9	42.9	14.3
48	Thoppayil	45.8	45.8	8.3
49	Chakkorath Kulam	72.7	27.3	0.0
50	Westhill	85.2	14.8	0.0
51	Varaykal	80.0	20.0	0.0
	Averages	**68.0**	**30.5**	**1.5**

Source: Field Survey, 2010.

Out of total household surveyed 813 (68.0 per cent) are R.C.C houses, 378 (30.5 per cent) are R.B.C houses and 20 (1.5 per cent) are muddy or thatched. The wards of Chevayur (91.3%), Nadakkavu (88.9%), Silver Hills (88%) and Cevarambalam (87%) are showing maximum and the wards of Vellayil South (40%) Koya valappu (36.4%), Payyanakkal (35.7%) Big Bazzar (33.3%) and Pallikkandi (30%) are showing least concentration of R.C.C houses. The wards Chakkumkadavu (72%), Idiangara (67.9%), Big Bazaar (66.7%) and Pallikkandi (56.7%) are showing maximum concentration of R.B.C houses and the wards of Vellayil North (14.3%) and Pallikkandi (13.3%) showing maximum concentration of the muddy and thatched houses.

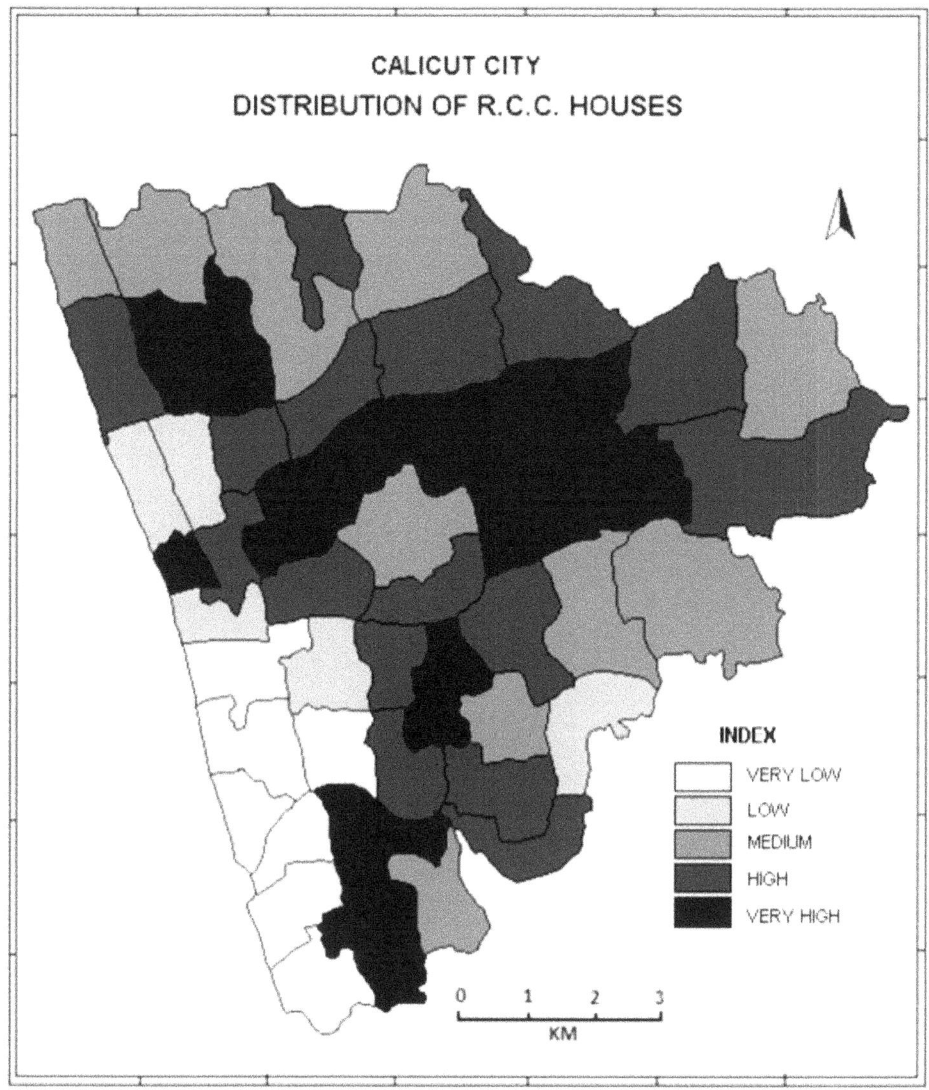

Figure: 5.3 Calicut City Distribution of R. C. C. House

5.7 Subscription of Periodicals

The newspaper is the most powerful of all the organs of expression of the news and views about men and the things. Newspapers keep us in touch with the current world affairs and extend the bounds of our knowledge and make us feel that we are a part of a living world. Newspapers are also an important means of communication between the government and the people. A newspaper is a daily magazine that gives the detailed information of all the events that is happening in the world. In democratic countries news papers have an important role in forming public opinion since they discuss on current events and condemn or admire the conduct of the government. The problems that face the country and the different possible ways of solving those problems are also a routine subject matter of the magazines and news papers. Thus, periodicals give awareness and education to the public and enable the citizen to have their own opinion on public affairs.

Table: 5.7 Subscription of periodicals in Calicut city in Percentages.

Ward No.	Ward Name	Daily	Weekly	Monthly
1	New Bazar	80.0	12.0	8.0
2	Edakkat	87.0	30.4	17.4
3	Easthill	84.0	24.0	32.0
4	Kuruvisseri	96.0	52.0	44.0
5	Malaparamba	81.8	31.8	9.1
6	Vengeri	87.0	26.1	13.0
7	Kannadikkal	73.1	7.7	19.2
8	Paroppadi	85.2	25.9	14.8
9	Chevarambalam	82.6	17.4	17.4
10	Kudilthodu	84.0	36.0	20.0
11	Cheveyur	95.7	34.8	39.1
12	Silverhills	92.0	40.0	32.0
13	Poolakkadavu	87.5	20.8	12.5
14	Moozhikkal	73.9	17.4	17.4
15	Chelavoor	73.9	17.4	13.0
16	Mayanadu	73.9	13.0	21.7
17	Kovoor	77.8	14.8	18.5
18	Nellikkode	87.0	30.4	21.7
19	Potamal	85.0	15.0	15.0
20	Kommeri	84.6	23.1	7.7
21	Pokkunnu	76.0	24.0	16.0
22	Mankavu	84.0	28.0	12.0
23	Kinasseri	90.9	36.4	22.7
24	Thiruvannur	95.7	39.1	13.0
25	Kallai	95.5	31.8	27.3
26	panniyankara	85.7	52.4	38.1
27	Meenchanda	92.3	34.6	11.5
28	Koya Valappu	50.0	9.1	4.5
29	Payyanakkal	46.4	14.3	7.1
30	Chakkumkadavu	28.0	8.0	0.0
31	Pallikkandi	40.0	10.0	3.3
32	Idiangara	64.3	17.9	7.1

Contd...

Table: 5.7 Contd...

Ward No.	Ward Name	Daily	Weekly	Monthly
33	Chalappuram	81.8	22.7	13.6
34	Azhchavattam	77.3	31.8	13.6
35	Kuthiravattam	91.3	34.8	39.1
36	Puthiyara	84.2	21.1	15.8
37	Palayam	61.5	11.5	3.8
38	Big Bazar	75.0	25.0	12.5
39	Vellayil South	60.0	8.0	4.0
40	Thiruthiyad	77.3	22.7	9.1
41	Kotooli South	81.0	23.8	14.3
42	Kotooli North	78.3	30.4	8.7
43	Civil Station	87.5	45.8	29.2
44	Eranchippalam	87.0	30.4	26.1
45	Karaparamba	85.0	30.0	20.0
46	Nadakkavu	88.9	27.8	11.1
47	Vellayil North	53.6	10.7	14.3
48	Thoppayil	62.5	12.5	4.2
49	Chakkorath Kulam	86.4	31.8	22.7
50	Westhill	92.6	14.8	7.4
51	Varaykal	80.0	20.0	30.0
	Averages	**78.7**	**24.5**	**16.8**

Source: Field Survey, 2010

Out of 1211 houses surveyed 945 (78.7%) households subscribe any news papers, 293 (24.5%) households any weekly magazines and 200 (16.8%) households any monthly magazines in their houses. The wards of Kuruvisseri (96%), Thiruvannur (95.7%), Chevayur (95.7%), kallai (95.5%), Westhill (92.6%), Meenchanda (92.3%), Silver hills (92%), Kuthiravattom (91.3%) and Kinasseri (90.9%) are leading in newspaper subscription and the wards of Chakkumkadavu (28%), Pallikkandi (40%), Payyanakkal (46.4%) and Koya valappu (50%) are least newspaper subscribing wards. The magazines subscriptions are more or less following the newspaper subscribing habits of the wards. Newspaper and magazine subscription also reflects the educational attainment and social awareness of the area and households.

5.8 Household Security Feeling

Most people don't want to live in a place where they feel that something awful has happened in the form of crime and this will decrease the liveability of any neighbourhood. Family security and safety is very important as crime and aggression is increasing day by day throughout the world. Everyone wants to protect themselves and their family from the increasing threats. We would do almost anything for the sake of our family including protecting them from known danger and harm. This can

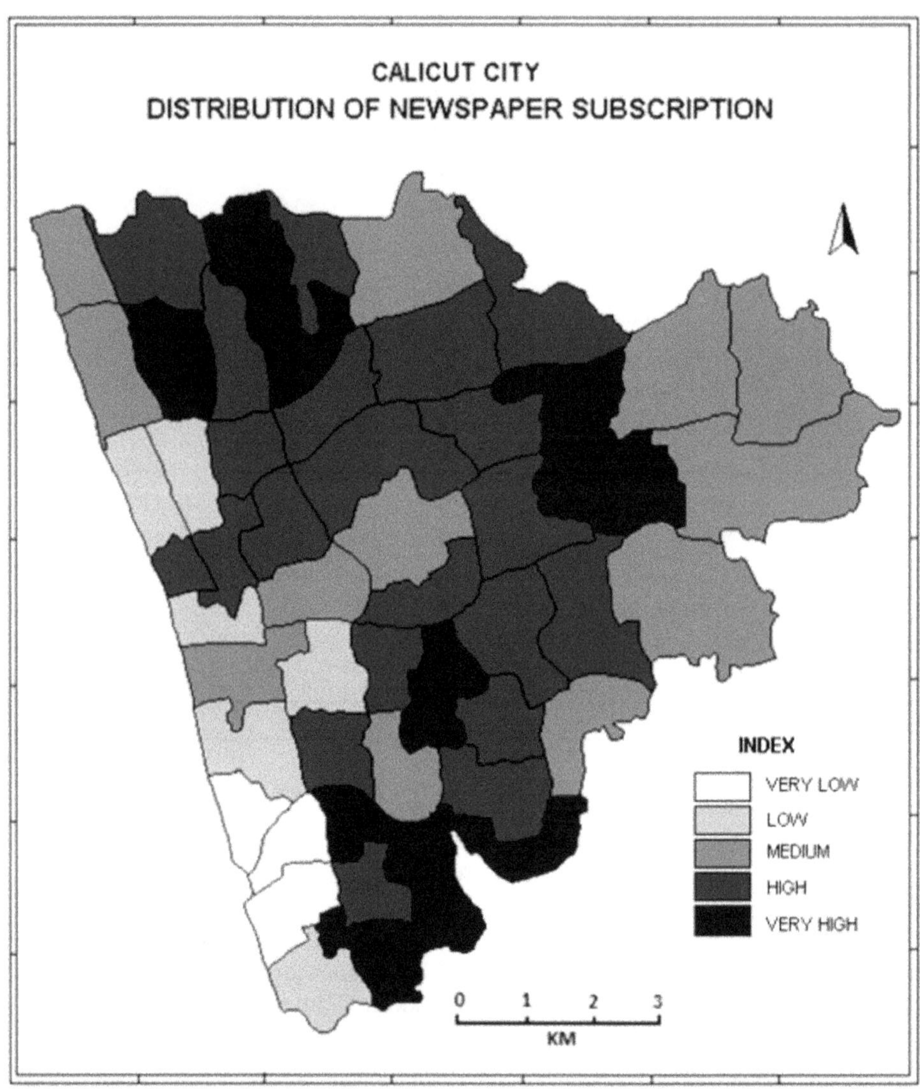

Figure: 5.4 Calicut City Distribution of Newspaper Subscription

be difficult because one cannot be with his/her children and kin at all times and lives in an open society where they can be exposed to predators and violent criminals. So the only possible counter measure for this Problem to create a safe environment in the society and locality, or move such places where security condition is prevailing. Thus all the people wish to have peaceful atmosphere in their locality or tend to move those places where he seldom despairs about the safety of the self and his family.

Table: 5.8 Household Level Security Feeling in Calicut City in Percentage.

Ward No.	Ward Name	Secure	Insecure
1	New Bazar	96.0	4.0
2	Edakkat	95.7	4.3
3	Easthill	92.0	8.0
4	Kuruvisseri	96.0	4.0
5	Malaparamba	95.5	4.5
6	Vengeri	91.3	8.7
7	Kannadikkal	100.0	0.0
8	Paroppadi	100.0	0.0
9	Chevarambalam	100.0	0.0
10	Kudilthodu	96.0	4.0
11	Cheveyur	95.7	4.3
12	Silverhills	96.0	4.0
13	Poolakkadavu	100.0	0.0
14	Moozhikkal	100.0	0.0
15	Chelavoor	95.7	4.3
16	Mayanadu	95.7	4.3
17	Kovoor	100.0	0.0
18	Nellikkode	95.7	4.3
19	Potamal	90.0	10.0
20	Kommeri	96.2	3.8
21	Pokkunnu	96.0	4.0
22	Mankavu	96.0	4.0
23	Kinasseri	95.5	4.5
24	Thiruvannur	95.7	4.3
25	Kallai	95.5	4.5
26	Panniyankara	95.2	4.8
27	Meenchanda	100.0	0.0
28	Koya Valappu	81.8	18.2
29	Payyanakkal	82.1	17.9
30	Chakkumkadavu	80.0	20.0
31	Pallikkandi	83.3	16.7
32	Idiangara	96.4	3.6
33	Chalappuram	86.4	13.6
34	Azhchavattam	95.5	4.5
35	Kuthiravattam	95.7	4.3
36	Puthiyara	94.7	5.3
37	Palayam	80.8	19.2
38	Big Bazar	66.7	33.3

Contd...

Contd...

Table: 5.8 Contd...

Ward No.	Ward Name	Secure	Insecure
39	Vellayil South	72.0	28
40	Thiruthiyad	90.9	9.1
41	Kotooli South	95.2	4.8
42	Kotooli North	100.0	0.0
43	Civil Station	100.0	0.0
44	Eranchippalam	91.3	8.7
45	Karaparamba	95.0	5.0
46	Nadakkavu	94.4	5.6
47	Vellayil North	75.0	25.0
48	Thoppayil	83.3	16.7
49	Chakkorath Kulam	86.4	13.6
50	Westhill	92.6	7.4
51	Varaykal	95.0	5.0
	Average	**92.5**	**7.5**

Source: Field Survey, 2010.

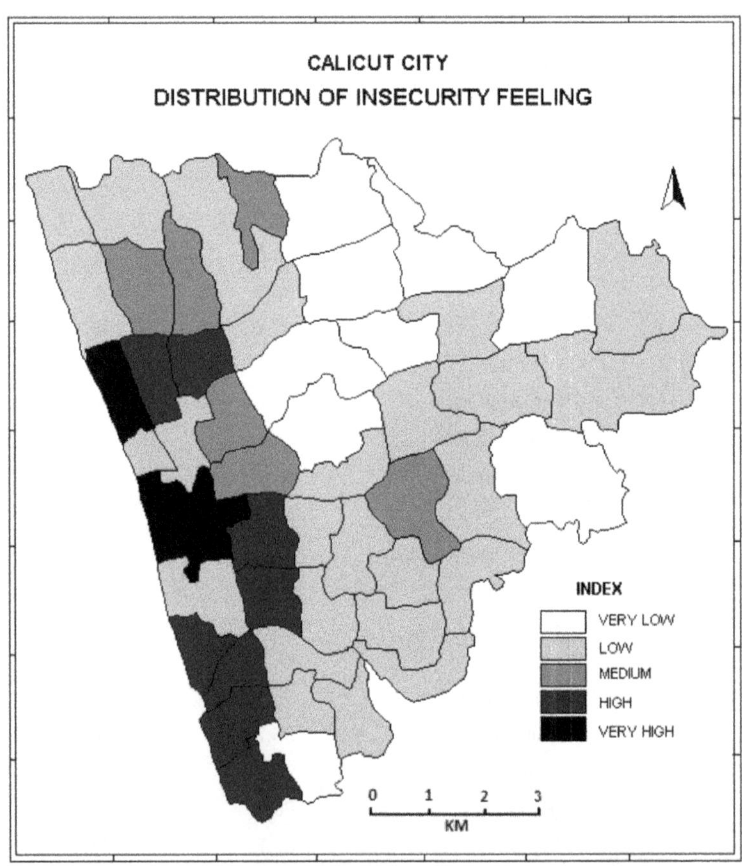

Figure:5.5 Calicut City Distribution of Insecurity Feeling

Out of 1211 only 93 (7.5%) households feel they are not secure in both the neighbourhood and in the house. And rest1118 (92.5%) households feel that there is no threat for their life and property in their neighbourhood. The wards like Kanndikkal, Paropadi, Cheverambalam, Poolakkadavu, moozhikkal, Kovoor, Meenchanda, Kotoli North and Civil station recorded hundred percentage of the households feeling secure. The households of the wards Big bazaar (33.3%), Vellayil south (28%), Vellayil north (25%), Chakumkadavu (20%), Palayam (19.2%), Koyavalappu (18.2%), Payyanakkal (17.9%), Pallikkandi (16.7%) and Thoppayil (16.7%) feel insecure for their life or property due to several reasons like the weakness of their houses, congestion and unfamiliarity in the neighbourhood.

5.9 Migration

Migration is the general movement of the people from one place to another either permanently of temporarily. Migration has important effects on economy, society, and politics and has impacts on areas such as housing, education, labour markets, population growth, and social cohesion. Urban regions usually attract migrants because of the availability of jobs and the presence of other facilities for migrant population. Around the world people move in and out of places every day, and it occurred throughout human history. Their patterns of movement reflect the conditions of an ever-changing world and, in turn, impact the cultural landscapes of the places they leave and the places they settled. These imprints on a region include its ethnic make-up, traditions, architectural styles, local food, music, clothes, and other cultural marks. Thus, an essential part of understanding a region is its migration story.

In the case of Calicut city internal migration is more and the immigration from outer region is very less. The main reason is that generally *keralites* are opting to live in villages due to the availability of the basic facilities within the villages and the proximity of easy access to the towns and cities with effective transporting system. Moreover, the nearby regions of the city have several urban pockets like Mavoor, Kunnamangalam, Ramanattukara, Feroke etc. where the urbanization process is in work at a faster rate than in the city. Linear pattern of settlement along the roads, railways and natural barriers like rivers and coastline are more prominent in Kerala than the settlement clusters. The negative population growth rate recorded by the city (-1.02) in last decade is notable in this respect. Here the migration is calculated within the time span of 20 years, the migrants before that considered as residents of the city.

Table: 5.9 Inborn and Migrant Population in Calicut City in Percentages

Ward No.	Ward Name	Inborn	Migrants	
			Internal	External
1	New Bazar	80.0	12.0	8.0
2	Edakkat	78.3	13.0	8.7
3	Easthill	72.0	12.0	16.0
4	Kuruvisseri	68.0	20.0	12.0
5	Malaparamba	68.2	18.2	13.6
6	Vengeri	73.9	8.7	17.4
7	Kannadikkal	73.1	15.4	11.5
8	Paroppadi	81.5	3.7	14.8
9	Chevarambalam	82.6	8.7	8.7
10	Kudilthodu	72.0	8.0	20.0
11	Cheveyur	78.3	8.7	13.0
12	Silverhills	76.0	12.0	12.0
13	Poolakkadavu	87.5	8.3	4.2
14	Moozhikkal	82.6	8.7	8.7
15	Chelavoor	82.6	8.7	8.7
16	Mayanadu	87.0	4.3	8.7
17	Kovoor	77.8	7.4	14.8
18	Nellikkode	73.9	8.7	17.4
19	Potamal	85.0	5.0	10.0
20	Kommeri	80.8	7.7	11.5
21	Pokkunnu	92.0	0.0	8.0
22	Mankavu	88.0	0.0	12.0
23	Kinasseri	81.8	4.5	13.6
24	Thiruvannur	95.7	0.0	4.3
25	Kallai	81.8	9.1	9.1
26	Panniyankara	81.0	4.8	14.3
27	Meenchanda	88.5	7.7	3.8
28	Koya Valappu	95.5	4.5	0.0
29	Payyanakkal	89.3	3.6	7.1
30	Chakkumkadavu	100.0	0.0	0.0
31	Pallikkandi	93.3	6.7	0.0
32	Idiangara	96.4	3.6	0.0
33	Chalappuram	86.4	4.5	9.1
34	Azhchavattam	81.8	13.6	4.5
35	Kuthiravattam	82.6	4.3	13.0
36	Puthiyara	94.7	0.0	5.3

Contd.....

Table: 5.9 Contd...

Ward No.	Ward Name	Inborn	Migrants	
			Internal	External
37	Palayam	69.2	7.7	23.1
38	Big Bazar	66.7	4.2	29.2
39	Vellayil South	92.0	4.0	4.0
40	Thiruthiyad	86.4	9.1	4.5
41	Kotooli South	81.0	9.5	9.5
42	Kotooli North	82.6	4.3	13.0
43	Civil Station	79.2	8.3	12.5
44	Eranchippalam	82.6	8.7	8.7
45	Karaparamba	90.0	10.0	0.0
46	Nadakkavu	94.4	5.6	0.0
47	Vellayil North	78.6	7.1	14.3
48	Thoppayil	79.2	8.3	12.5
49	Chakkorath Kulam	86.4	4.5	9.1
50	Westhill	85.2	7.4	7.4
51	Varaykal	90.0	5.0	5.0
	Averages	**83.0**	**7.3**	**9.7**

Source: Field Survey, 2010.

Out of 1211 houses surveyed 1004(83%) are inborn in the city where they residing now, 207 (17%) are migrated either from outside the city or other parts of the same city. 88 (7.3%) households from inside the city and 119 (9.7%) from outside the city. Maximum percentage of inborn population is seen in Chakkumkadavu (100%), Idiangara (96.4%), Thiruvannur (95.7%), Koya valappu (95.5%), Puthiyara (94.7%), Nadakkavu (94.4%) and maximum percentage of migrants are seen in Big bazaar (33.3%), Kuruvusseri (32%), Malapramba (31.8%) and Palayam (30.8%).

6

STATUS OF URBAN ENVIRONMENT AND SOCIAL WELLBEING IN CALICUT CITY

Human wellbeing can be simply identified as the ability and the opportunity of individuals or groups to live in a manner that they appreciate. People's ability to perform the lives that they appreciate is shaped by a wide range of contributory rights. Human wellbeing covers personal and environmental security, access to materials for a standard life, sound health and good social relations and all those which are associated to better life, and the freedom to make choices and take action.

Urban environment and social wellbeing differs from one city to another even within the cities. These differences in environment and wellbeing originate from historical state of affairs of a town or city passes through, and different socioeconomic, cultural, administrational, facilities, amenities and security conditions. Existing facilities, the nature and the sense of its uses moulds the specific urban environment in localities along with the influence of natural environment. The social environment or wellbeing related with the attitude and nature of social relations and social contact aggregated with the education and social commitment.

Urban environment is the aggregates of the subsystems of the city or towns like residential patterns, housing condition in terms of per capita indoor space, open space and the materials used in the construction, existence and the efficiency of transportation networks, presence and permanency of greenery, proper working condition of the sewerage network and street lights, efficiency of municipal distribution systems of drinking water and cooking gas, presence of leisure facilities like sports complexes and theatres, collection, processing and disposal of solid waste.

6.1 Multivariate Analysis of Urban Environment and Social Wellbeing

The characteristics of the urban environment and the social of households and locality are unlike in different parts of the city. But there must be an array of common characteristics on which household, locality or ward can be differentiated. Social area analysis is the most commonly used methodology for this purpose, but recently researchers use different form of multivariate analysis like linkage analysis, cluster analysis, principal component analysis and common factor analysis to explore the organization of environment and of the city space. It is important to discuss the dimensions of spatial differentiation of households observed by the researchers in quality of life or analyses in general, and those attributes used by the author for this analysis in particular.

6.1.1 Housing Structure

The immediate housing environment and the neighbourhood represent an everyday-landscape, which can either support or limit the physical, mental, and social wellbeing of the residents. The housing environment has been acknowledged as one of the main settings that affect human health. The housing and living conditions are influencing residential environment in many ways. Indoor air quality, home safety, noise, humidity, indoor temperatures, presence of asbestos, lead, radon and volatile organic compounds, lack of hygiene and sanitation, and crowding are some of the most relevant possible health threats to be found in dwellings. Physical, mental, and social health is affected by the living conditions, but no straightforward quantifying mechanisms have yet established. The quality of housing conditions plays a decisive role in the health status of the residents. Representing the spatial point of reference for each individual, the home also has a broad influence on the psychosocial and mental wellbeing by providing the basis for place attachment and identity as well as a last refuge from daily life. (Bonnefoy 2007, p.411-412)

A dwelling is defined as a holding space, a physical and psychological envelope within which intimacy will appear and develop and where each and every individual will find an opportunity to be himself or herself. Thus, what was just a house will become a home. Integrity of body and mind are dependent upon this possibility of living in intimacy (Bonnefoy 2007, p. 419). Even today there is not a commonly agreed definition of 'healthy housing', and there are still major gaps in the knowledge that, how housing conditions may affect health and which mitigation strategies may show the best results.

At some extend the building materials used in the construction of the houses like reinforced cement concrete (RCC), reinforced brick concrete (RBC), wooden and thatched are have direct effect on the human being in both physically and mentally. If we consider the mental health, one of the most influencing factors may be the indoor space and open space. Per capita indoor space inside and open space outside the dwelling affect the individual, family and society. Above all, the ownership of houses also contributes to the environment and wellbeing especially in the sense of housing security.

6.1.2 Amenities and Facilities

Amenities and facilities can be consider as all those play a part in the better living of the individual, family and society. Facility need of the people will change according to time and space. For example in hot countries cooler and air conditioner are important and essential facility but for those living in cold countries these are not a necessary but the blower and heater may the basic and inevitable need in their daily life. Amenities are any tangible or intangible benefits of a property, especially those that increase its attractiveness or value or that contribute to its comfort or convenience. Some people also use the word "amenity" for lavatory, bathroom and personal hygiene facilities.

Access to safe water and sanitation is a human right. More than a billion people in the developing world lack safe drinking water. Nearly three billion people live without access to adequate sanitation systems necessary to reduce exposure to water-related diseases. The failure of the international aid community, nations, and local organizations to satisfy these basic human needs has led to substantial, unnecessary, and preventable human suffering.

Telephone/ mobile and electricity are the basic necessity of the mankind in recent years and without these the life will be a misery. Moreover electronic and printing Medias like television computer and newspapers are became the charisma of the modern civilized population. They keep them updated with the world affairs. Other innumerable facilities of different dimensions are also play important role in the life of city and society. Open spaces, sports and recreational facilities have also a vital role to play in promoting healthy living and preventing illness, and in the social development of children of all ages through play, sporting activities and interaction with others.

6.1.3 Demography and Ethnicity

Family wellbeing lies at the core of civil society. It recognizes the particular role that families play in society, and that certain functions provided by families cannot be provided in any other way by the state or any other social institutions. The family loves, nurtures and protects its members- particularly children, is responsible for their development, education and care, provides a moral framework within which the family and its members operate and connect with wider society, and ensures the intergenerational transmission of cultural history and norms.

Family wellbeing differs from individual wellbeing that it also refers to the overall wellbeing of a social form or structure which is different to the sum of the wellbeing of its individual members. Families exist as individual units, and also as participants within the community and the broader social, economic and cultural environments in which they exist. All members of families can be safe, can grow and develop to lead healthy, happy, stable, fulfilling, socially engaged lives, and contribute to the wellbeing of the family and society

Religion provide an important basis for ideas about wellbeing specifying through teaching and practice what it mean to live well, as an individual and as a community. It is also widely understood as a source of wellbeing for its adherents, providing comfort in times of trouble, offering a frame work of meaning to make sense of life's variations, and providing a community that gives social support and confers identity through a sense of belonging (White et al. 2010, p.1). Composition of ethnic and religion groups like Hindu, Muslims and Christians are also have impact on social environment and wellbeing. The value system and the preferences are differing from one group to another. One group may give well attention to family organization, moral values and religious thought, at the same time other group may not have much care in these but will give good attention to education and the source of income.

Size of households and number of children contribute to overall mental and physical growth of individual and hence the society. Per capita time availability with their parents and elder ones is depends on the number of children and it is very influential in their educational and personnel performance. Type of family is also has well defined effect in the formation of well society. Considering the debatable state of the matter in the modern progressive society both of the nuclear and joint families have their own advantages and shortcomings. So any analysis which related to the social environment cannot avoid these two systems of family organization.

The composition of general and backward category are also have important role in moulding the society. It may cause the social stratification which is not a sign of the well society. Commonly the general category have more affluence in the case of income, wealth, social status and other dimensions of the wellbeing than that of the backward classes not only by worthiness but also by some historical factors like slavery and labour compartmentalization. The back ward sector will take time to break away from this social inequality and to capture the social capillarity comparable to the higher classes.

6.1.4 Education and Income

A sound education provides the foundations for the future ability to make choices regarding one's occupation, thereby giving a greater influence and control over his future levels of income, where he will live, and also over the various factors that influence his health and wellbeing across the lifespan. A high level of education equips individuals with the skills to cope with day to day challenges, and further enables individuals to participate more actively within the employment market, the economic market, and their communities. The level of education attained by individuals is inseparably linked to the social gradient. The divide between those who have access to information technology and those who don't is undoubtedly an inequitable situation (Cannon 2008, p.9).

Knowledge is power, so to be educated is to be empowered. Education allows human being to surpass poverty and ignorance, to become independent decision-making members of their family and society. Education should help the people to discover lasting values which help them to break down our national and social barriers, instead of emphasizing them. Education is must for each and every person. Generally the illiterate persons are poor and cannot compete with educated people in the sphere of health, govt. service and even in business life.

The individual's personal income effect on health and wellbeing and those have greater level of personal income enjoys better access to health care, nutrition and often live longer than people on the lower income. Within the upper levels of the social gradient other social outcomes such as education, literacy and coping behaviours are overrepresented (Cannon 2008 p.7). There is a positive correlation between individuals income and his or her subjective wellbeing because it aids individuals in meeting universal human needs such as good health, nutrition, and comfortable housing, food, safety, they prone to have greater subjective wellbeing, regardless of social comparison and so forth (Diener et al. 1993, pp.195-197)

6.1.5 Security and Migration

A perception of safety in a community is a keystone factor to healthy and vibrant communities, in the same way that crime is an indicator of some level of societal dysfunction. It has been suggested that in communities where there are shared values and norms, and where good informal social networks operate in neighbourhoods, the community may enjoy lower levels of crime and the higher level of security feeling. In addition, a perception of safety in the community is important to encourage people to feel confident in participating in activities in their community such as sport, recreation and cultural activities (Australian Bureau of Statistics 2002, p. 14)

The problem of human security is a major element in the understanding of wellbeing. The idea of security is inseparably associated with law and order and rights but the social conditions are very important to this, (Geof 2007, P.109). High crime rates can also diminish social resources such as community trust, confidence and freedom, and an overall climate of fear may overwhelm or replace the spirit of cooperation and participation in community life. The emotional and physical safety is key factor to the healthy development and wellbeing of the individual and society.

6.2 Variables Selected

The characteristics of the urban environment and the social wellbeing of households and locality are unlike in different parts of the city. But there must be an array of common characteristics on which household, locality or ward can be differentiated. Social area analysis is the most commonly used methodology for this purpose, but recently researchers use different form of multivariate analysis like linkage analysis, cluster analysis, principal component analysis and common factor analysis to explore the organization of environment and wellbeing of the city space. It is important to discuss the dimensions of spatial differentiation of households observed by the researchers in cross cultural analyses in general and used by the author for this particular analysis.

In order to draw out the spatial variation of the social wellbeing in Calicut city forty-nine variables are selected which the author thinks are indicative of urban environment and the social wellbeing. These variables reflect the process, states and facts of the social environment of the city. They relates to five sets which includes all significant housing, material infrastructure and other elements of environment of the wards. These are (1) Housing structure, (2) Amenities and facilities, (3) Demographic and family, (4) Education and income and (5) Security feeling and migration.

Housing structure include ten variables represents the quality of housing of the wards which, to a great extent, reflects the general wellbeing of the city with special reference to the general environment, housing environment and indoor environment. The set of amenities and facilities is composed of sixteen variables which pertain to the sanitation condition, availability of drinking water, telephone and electricity connection, possession of television, computer and car, which have a paramount role in the wellbeing of household and family. Some of them are basic necessities and others are related to the leisure. The variable of demographic and ethnic structure incorporate the size of household, type of family, ethnic composition, and the category in social hierarchy which is important to represent social organization. The variable set education and income status include the literacy rate, education of family head, and the different income groups which have a paramount role in the assessment of a society as it control the aspiration and expectation which regulate the perception of the life and the mentality to avail the accessible facilities. The last variable set of security feeling and the migration status connect to six variables, first three of which indicate security feeling and the rest is the migration status.

The selection of variables is an attempt to portray all the aspects of the quality of urban environment and the social wellbeing in context of Indian cities. The author understands that, the nominated variables reflect the actual condition of the urban environment and social wellbeing. However this selection of variables is in no way is complete or comprehensive as a common measure. Due to the limitation imposed by the availability of the information at ward level and the available computer programme of factor analysis restricts the number of variables in relation to the number of observation, a large amount of variables compelled to omit. However some variables are left out from the selection seeing that some other important variable or set of variables are more or less explaining them. In some cases intermediate and in other extreme variables are ignored. By tolerating all these, it is felt that selected variables reasonably represent the problem which is under investigation.

6.3 Factor Structure

Forty-nine variables (indicators) representing the socioeconomic, infrastructural, demographic and physical aspects were taken for factor analysis. These variables has been clustered under such five sets such as housing structure, amenities and facilities, demographic and family structure, education and income structure and the security feeling and migration structure. Factor analysis of forty-nine variables which determine the urban environment and social wellbeing in Calicut city has yielded five significant factors which can be broadly named as housing and material status, family status, education status, security status, and migration status. These five factors together explain 83.95 per cent of the total variance in social wellbeing. The contribution of every factor varies from each other. The first factor contributes 29.03

per cent, the second one have a share of 18.35, while the others have 17.22 per cent, 10.41 per cent and 8.44 per cent respectively to the urban environment and the social wellbeing making a total of 83.95 per cent.

Table: 6.1 Factor structure of multivariate analysis

Sl.No.	*Factors*	*Percentage of total variance explained in social well being*
1	Material and housing status	29.03
2	Family status	18.85
3	Education status	17.22
4	Security status	10.41
5	Migration status	08.44
Percentage of total variance explained by five factors		**83.95**

The loadings of forty-nine variables reveal that some of them are important in explaining the material status and housing condition, and some others demographic status, education status, security status and migration status. But there are some indicators which are important in more than one factor, for example the variable 'household with open space' and 'household without open space' are important in factor 1 that is material and housing status as well as in factor 4 that is security status. Moreover some other variables like percentage of Christian population, RBC houses are not a strong association with any factor, but only a marginal importance.

6.3.1 Factor- 1 Material and Housing Status

Factor 1 which explains 29.03 per cent to the total variance of social wellbeing in Calicut city can be identified as the dimension of the material and housing condition. The nature of this factor is clearly defined by the high loadings of seven housing structure variables, fourteen amenities and facilities variables and three education and income variables. These all categories of variables are related to the material and housing status and hence the social wellbeing in one or the other ways and are be interpreted as the essential indicants of the material and housing status.

Table: 6.2 Material and Housing Status of Calicut City

S. No	*Variables*	*Factor Loading*
	HOUSING STRUCTURE	
1	R.C.C. Houses (per 100)	**0.5522**
2	R.B.C Houses (per 100)	**-0.4526**
3	Huts/ thatched Houses (per 100)	**-0.8143**
4	Owned houses (per 100)	0.0235
5	Rented Houses (per 100)	-0.0235
6	Houses with open space	**0.5102**
7	Houses without open space	**-0.5091**
8	Houses with 1-2 bedrooms (per 100)	**-0.7123**
9	Houses with 3-4 Bedrooms (per 100)	0.1236

Contd....

Table: 6.2 Contd...

S. No	*Variables*	*Factor Loading*
	HOUSING STRUCTURE	
10	Houses with above 4 bedrooms (per 100)	**0.4693**
	AMENITIES AND FACILITIES	
11	Houses with sanitation Facilities	**0.8544**
12	Houses without Sanitation Facilities	**-0.8554**
13	Houses with Running Water	**0.7463**
14	Houses without running Water	**-0.7475**
15	Houses with telephone/ mobile	**0.9072**
16	Houses with no telephone/mobile	**-0.9065**
17	Houses With Electricity	**0.8855**
18	Houses without Electricity	**-0.8864**
19	Household with Bank Account	**0.7679**
20	Household without Bank Account	**-0.7697**
21	Houses Subscribing News Paper (per 100)	**0.6756**
22	Houses with T.V	**0.6938**
23	Houses without T.V	**-0.6942**
24	Houses with Computer	0.3351
25	Houses without Computer	-0.3329
26	Houses with Car (per 100)	**0.5800**
	DEMOGRAPHY AND ETHINICITY	
27	Small household (per 100)	0.0051
28	Medium household (per 100)	0.3014
29	Large household (per 100)	-0.2673
30	Nuclear Family (per 100)	0.2450
31	Joint Family (per 100)	-0.2443
32	Hindu population (per 100)	0.3545
33	Muslim population (per 100)	-0.3510
34	Christian population(per 100)	0.0972
35	General Category (per 100)	0.2440
36	Backward Communities (per 100)	-0.2769
37	Work Participation Rate	**0.4600**
	EDUCATION AND INCOME	
38	Literacy Rate	**0.6423**
39	Family Head having High School Education	-0.4065
40	Family Head Having diploma/ graduation	0.4066
41	Low Income Household	**-0.5755**
42	Medium income Household	**0.6154**
43	High Income Household	0.2979

Contd.....

Table: 6.2 Contd...

S. No	Variables	Factor Loading
	SECURITY FEELING AND MIGRATION	
44	Houses with High Security Feeling	0.0244
45	Houses with a feeling of Secure	0.2151
46	Houses with Insecure Feeling	-0.2493
47	Inborn Households	-0.2995
48	Household Migrated within the city	-0.0399
49	Household migrated from outside city	**0.4515**

The positively loaded variables are associated with higher status and negatively loaded variables are denoting the lower status in material and housing condition. Factor -1 determined largely by twenty-five out of forty-nine variables, seven of which related to housing structure namely houses with only 1-2 bedrooms, household with above four bedrooms, houses with open space, houses without open space, houses of RCC, houses of RBC and the percentage of huts and thatched houses. Fourteen related to the amenities and facilities namely- houses with sanitation facilities, houses without sanitation facilities, houses with running water connection, houses without running water connection, houses with telephone/ mobile connection, houses without telephone/ mobile connection, houses with electricity, houses without electricity, houses with bank account, houses without bank account, houses subscribing news paper, household with television, household without television and houses with car. Other three related to education and income status namely- literacy rate, low income household and middle income household and one related to demography and family that is work participation rate.

Among these variables, highest positive loading is recorded by the telephone connection (0.9072) followed by electricity connection (0.8855), sanitation facility (0.8545), bank account (0.7679), running water connection (0.7464), possession of television (0.6939), subscription of news paper (0.6756), literacy rate (0.6423), medium income household (0.6154), car ownership (0.5800), R.C.C. houses (0.5522) open space (0.5102), houses having above four bedrooms (0.4693) and work participation rate (0.4601) are loaded positively. Variables of positive loading denote that they favour the urban environment and the social wellbeing in Calicut city and the high loadings imply that they are very important to explain social wellbeing and urban environment.

Some variables are loading high but negatively on this dimension of social wellbeing. The highest negative loading is registered by the houses without telephone/ mobile connection (-0.9065), followed by houses without electricity (-0.8865), houses without sanitation facilities (-0.8554), percentage of huts and thatched houses (-0.8143), household without bank account (-0.7697), houses without running water (-0.7475), houses with only 1-2 bedrooms (-0.7123), houses without television (-0.6940), low income household (-0.5756), houses without open space (-0.5092) and RBC houses (-0. 4526). The variables which negatively load to material and hosing status denote that the high loading of these variables are inversely affecting the social wellbeing and urban environment of the city and society.

The presence of telephone connection, electricity and sanitation facilities are loading positively and the absence of them loads negatively. Telephone/mobile connection is very important to the social contact and other necessary communications in the modern society. In a society like the people of Calicut city in which the level of social contact and community involvement is much higher, it is very important to have electronic communication devices like telephone /mobile. The non existence of this facility in any household indicates either the extreme poverty or the tendency of withdrawal from the social contacts and community involvements, both of which are very chaotic to the urban social environment and wellbeing. Moreover for most of *keralites* native place and workplace are different, so they want this facility to keep contact with their kin and friends.

Electricity is an essential part of day to day life of modern society without which people cannot fulfil their household energy needs where they spent a lion share of their time of a day. It can be say one of the essential requirements of modern man and the inaccessibility to this reveals either the low economic condition or the illegitimacy of the dwellings or dwelling place which desperately affect the urban and social environment of a city.

Sanitation facility is also very important to any household that is why the existence of this loaded highly and positively and the absence is loaded highly and negatively. Every human being requires private places for their sanitation necessities. Most of the residence dwellings of the Calicut city have own toilets with septic tank and flushing facilities, in new residences it is inside the houses and old residences in outside houses, in some areas of slum like condition it is common for several families or not exists that is why the wellbeing of that areas recorded very poor in this regard.

Water is a basic necessity to all living organisms. Man uses water both for his biological and domestic needs. Accessibility to good water is very important to all households. In Calicut city most of the households uses their own private wells as source of water and a large number of households depend on corporation connection and some both. Even a considerable size of households depends on public water taps or other public sources like wells and ponds. It is very clear that the easy and cheap availability of domestic water leads to the wellbeing and the unavailability leads to ill-being of the urban environment and the social wellbeing.

Bank account and bank balance shows the prosperity of the individuals, households and hence the community and society of any locality especially in economic aspects. So higher number of bank accounts favouring and the lower numbers disservices the urban environment and the social wellbeing. Banks and banking are one of the inevitable services in the modern life. Now the paper currency is transforming to electronic currency or bank credits. Almost all transactions of higher amounts are being done through banks in recent years. So not being a part of this facility denotes the poor status of living.

Similarly positive loading of possession of television and subscription of news papers shows paramount importance in this factor. These two electronic and printing Medias keep the people in touch with the day to day development of socio cultural, economic, technological and political in regional, national and international dimensions. As an almost hundred per cent literate city, these two communication medias are playing very important role in the life of the people and contribute a good city life and social environment. At the same time the absence of television is loaded negatively which means the city life without the access to it will affect the material condition and hence the social environment and wellbeing.

Literacy rate is another important variable which loads positively. Education and literacy is an essential capital for the development in the modern era of information revolution. Calicut city has high literacy rate in comparison with other Indian cities due to the high facility in basic education sector, and able to compete with cities of developed nations if we ignore the literacy of aged population. Education brings the employment and good status of life that is an integral part of the wellbeing.

Medium income household and car ownership are also has positive loadings of marginal importance in the material and housing status. Income may be considered as the base of the material and housing dimension along with other variables, because money is essential for the attainment of all the requirement of the human being. More than half of the households of Calicut city come under the medium income group that is annual income between 2- 5 lacks. Car ownership is nothing but the reflection of the income and the degree of necessity.

Other marginally important positive loading are recorded by RCC houses, presence of open space and houses having more than 4 bedrooms which are totally related with housing space and construction materials. Space gives relaxation and enthusiasm to human beings and the congestion led to stress and disturbances. So the open space in the houses and working places are very important to the mental and psychological wellbeing of the individual and society. The housing pattern in Calicut city and Kerala allows enough open space in all sides of houses in general and in front of it specifically. The side walls of houses are widely separated with the neighbouring houses with at least 5-10 metres space between them. There is courtyard system around the dwelling as a mark of the tradition and culture of the society.

Building material and the indoor space are also very important in the housing status. RCC houses have comparatively bad conductivity of moisture and minimize the growth of minute organisms and pathogens at the floor and walls. The indoor space provide free and relax feeling while the congestion gives pressure that is why these variables are positively loaded in material and housing status. There is another problem of unavailability of traditional skilled labours to use the indigenous materials to construct the traditional houses.

On contrary to this the variable huts/ thatched houses and RBC houses loaded negatively to the factor material and housing status. RBC houses have a possibility to hold moisture in their tiles and other materials due to poor surface smoothness and they permit the life of micro-organisms and pathogens especially in the areas of tropical humid climate like Calicut, which adversely affects the human health and hence the wellbeing. The aforesaid vulnerability is very much higher in the case of huts/ thatched houses than the RBC houses. There is less security and relaxation feeling in these cases by reason of the possibility of easy entrance of robbers and thieves due to the defenceless materials of the roof and walls.

Low income group of household and the household having only 1-2 bedrooms are also loaded negatively on this factor. These two variables are mutually interrelated. Low income leads to the small houses, deteriorated quality of the life and less accessibility to the standard goods and services. Low per capita indoor space leads to the conjunction and inadequate living space inside the dwellings affect the productivity and the mental health of the people. Work participation rate is another positively but marginally loaded variable on this factor. Employment and work participation of any community will determine the development of the material status of the people living them.

Ongoing discussions point towards the material possession, housing structure, amenities and facilities, and education and income are the prime components of urban environment against which the social wellbeing and the liveability can be judged. The factor 'material and housing status' indicates that the variable set amenities and facilities is the most important to the urban environment and social wellbeing in Calicut city followed by the housing structure and the education and income.

Thus factor 1 based on its close association with the measure of housing structure, amenities and facilities, education and income and demography, appears to be a dimension of material and housing status as hypothesised in the social area analysis. The factors scores are the standardized measure of the wards on this factor have been divided in to five class interval: very high, high, medium, low and very low.

Fig: 6. 1 shows the spatial differentiation of the quality of urban environment and social wellbeing expressed by the dimension of the material and housing status. Out of 51 wards eight exhibit very high factor score that is these wards are comparatively excellent in material and housing condition, 21 wards scores high that means they are in good condition, 14 wards shows medium that is average, 5 wards records low that is below average and 3 wards tallies very low that is poor condition in the material and housing status.

Spatial pattern of the wards according to the material and housing status clearly organize the city in to four distinguished region with a few exceptions. The inner city comprising five wards showing very high score in the material and housing status, and the inner circle around it having high score, and the peripheral region with the average condition and the coastal line with poor and very poor condition in the same. The south eastern part of the city showing mixed condition.

The very high or comparatively excellent region in the material and housing status is situated mainly in the geometrical centre of the city comprising five wards namely Chevayur, Silverhills, Kudilthodu, Kotooli south, and Civil station. These areas are relatively new settlement lying in the vicinity of two important roads- Calicut-Wayanad road (NH) and Mavoor road. And other three wards coming under this category is located in the two extreme ends Thiruvannur in the south and Vengeri and west hill in north.

Almost all positively loaded variables which explain the material and housing status showing very high percentage in the centre of the city such as telephone connection, sanitation facilities, RCC houses, open space and the indoor space. These

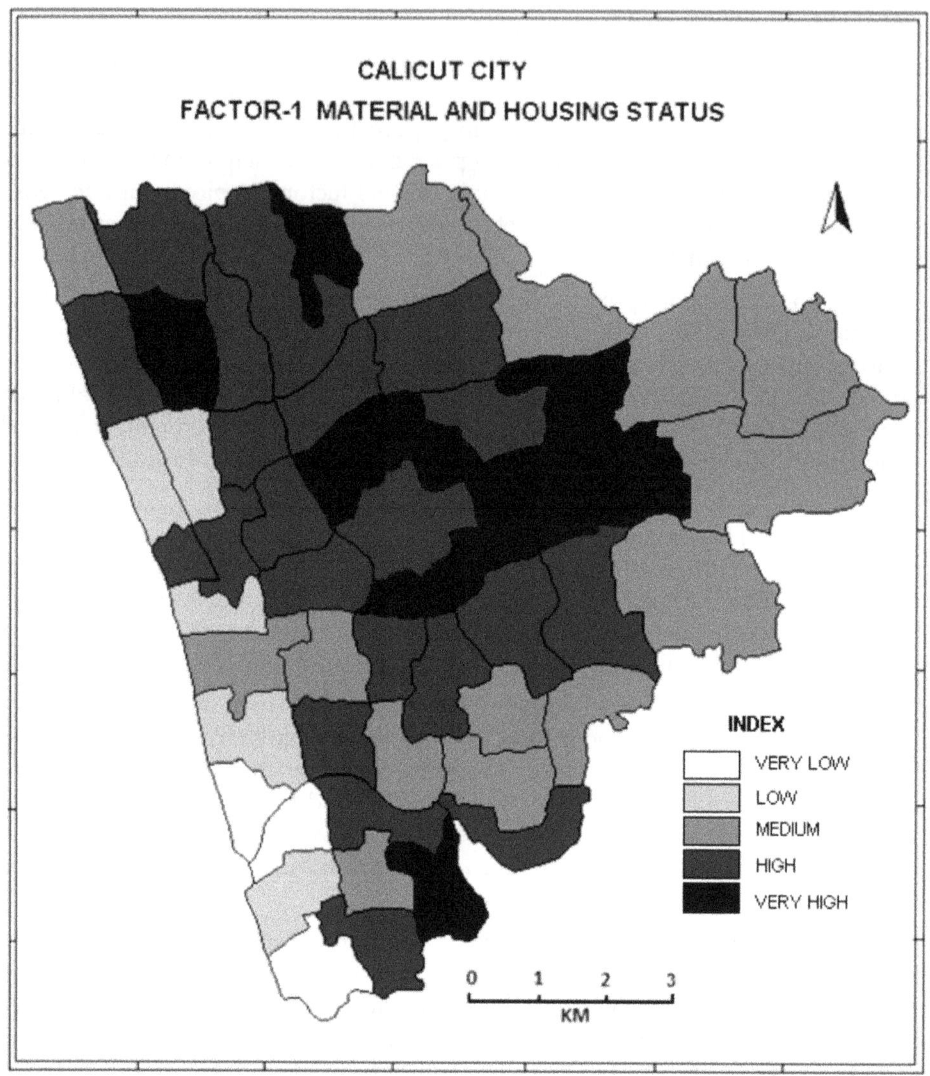

Figure: 6.1 Calicut City Factor-1 Material and Houseing Status

wards are exclusively residential areas and reside largely by government officers, and professionals due to the vicinity of Civil Station, Medical College, Law College and plenty of other government and private educational institutions. This is a slum free area and having much open spaces and greenery including the kotooli wetlands where mangrove forests and perennial water bodies are present. Most of the dwellings are of RCC, having enough indoor space and have a large number of facilities like running water and electricity connections etc. The ward Vengeri in the north is also a new residential area resembling the above wards in almost all facilities. The southern isolated ward Thiruvannur is something different from other wards. It is an old residential area of high caste Hindus having historical background and are the descendents of the former rulers of Calicut that is Zamorins and their relatives.

The region which registers the high factor scores mainly concentrated in the central and northern parts of the city almost adjacent to the very high region. This region comprises the wards Edakkat, East hill, Kuruvisseri, Malaparamba, Paroppadi, Chevarambalam, Nellikkode, potamal, Kuthiravattam, Puthiyara, Chalappuram, Thiruthiyad, Kotooli north, Eranchippalam, Karaparamba, Nadakkavu, Chakkorathkulam and varaikal. The three wards Kallai, meenchanda and Kinasseri, clustered in southern parts of the city.

Most of the wards coming under high category are more or less similar to that of the highest loaded region. The central region is a part of older city residential area which moving towards commercial centre and is densely resided. The material and housing status of this region with business men and professionals are very good but few pockets having slum like condition deteriorates the overall quality of life of the region. The open space and greenery is lesser in this region in comparison with very high region. The northern wards are characterised by a special features of mixed housing condition with the new and old houses of RCC and RBC.

The wards having medium factor score in the material and housing status make a crescent like appearance located in the eastern peripheral region and the central parts of the city as an outer covering of the high scored region. The region comprises the wards namely Kannadikkal, Poolakkadavu, Moozhikkal, Chelavoor, Mayanadu, Kovoor, Kommeri, Pokkunnu, Mankavu. Azhchavattom, Palayam and Big bazar. The two wards New Bazar- located away from the cluster in the north western corner and Panniyankara in the southern part are also coming under this group.

The medium scored peripheral region has suburb characteristics. There are plenty of private owned open spaces and greenery, the density of dwelling is relatively very less and they are depend on the city centre for their higher facilities like marketing, higher education and medical care etc. The ward Big bazaar and Palayam coming on the CBD of Calicut city and large portion of them occupied by business establishment and market place. Thus the area is no more suitable for the residential purpose, people moves towards periphery from this region and now characterized by rented or temporary residents of the workers of business establishment and markets.

Low and very low scored region in the material and housing condition is totally of coastal stretch. Payyanakkal, Kuttichira, Velleyil south, Velleyil north, Idiangara and Thoppayil coming under low factor scored wards. Koyavalappu, Cakkumkadavu and Pallikkandi are coming under very low scored wards. These wards are thickly populated, majority of the people are low income group and some of them are fishermen. Their houses are of very poor in quality with bricks and wooden or thatched, resembling the slums.

Housing condition of this area is very poor with more than half of dwellings are of RBC and even a considerable amount of huts/thatched houses. Accessibility of water is also a problem in this region that many households fetching their domestic water from the public sources. Houses are closely spaced, the open spaces are very limited and per capita indoor space is very less with large family size and small land holdings. This region is largely inhabited by low and medium income groups. The sanitation facilities are also vulnerable, a considerable people using the open land for their sanitation needs. Electricity connection is also absent in some dwellings due to the illegal occupancy. The most of the slum sites in the city is located in this region of coastal line. Among these Pallikkandi, Idiangara and Koyavalappu are the most vulnerable areas.

A comparison of this distribution of the material and housing standard suggests that this dimension of quality of urban environment and the social wellbeing is strongly correlated with the socio-economic and housing status of the population.

6.3.2 Factor- 2 Family Status

Factor 2, the second most important factor explaining 18.85 per cent of the total variance in the urban environment and the social wellbeing in Calicut city can be identified as the dimensions of the family status. The family status of social wellbeing is largely determined by twelve out of forty-nine variables, ten are related to the demography and ethnicity namely small households, large household, percentage of nuclear family, percentage of joint family, percentage of Hindu population, percentage of Muslim population, percentage of general category and the percentage of backward communities, percentage of Christian population, work participation rate and two from housing structure such as RCC houses and RBC houses (table 5.3).

Among the variable loading higher is the nuclear family (0.9034), followed by small household (0.8902), percentage of general category (0.8537), percentage of Hindu population (0.8075), work participation rate (0.4782), Christian population (0.4717) and RCC houses (0.4446) are compassionate of good urban environment and social wellbeing explained by the factor family status. The wards showing higher loading in these variables are have higher status of family and those loading marginally or imperceptibly have faint status in family.

The family is the basic unit of any society and most important social unit fostering good values and responsible living. The objectives of the family are best met when its members are healthy in physical, mental and social aspects. Advantage of a joint family as an important faction of the society since the genesis of mankind is that the human interaction and bonding between all family members proves to be a great base for an individual's growth. A person's constitution, mind, thought process and ethics get well shaped in such a situation. This system is still common in several parts of the worlds like India. The joint family system is on a gradual decline in many Asian countries due to urbanization, modernization and westernization -that emphasize individualism

Availability of stable environment, good behavioural stability, a strong sense of consistency that provides a strong foundation to their roots, better learning skills with more per capita availability of time with their parents, mentality of sharing responsibility and the concentrated physical and emotional support in nuclear family is very high as compared to joint families. A nuclear family have a advantages of the feeling of independence, Involvement of children, Emotional attachment between family members, self-confidence, chances of greater interaction and privacy which lead to stronger emotional bonds and financial security. The emotional fulfilment and security becomes equally comforting and apt for overall wellbeing. The concept of nuclear family started becoming quite common in post industrial revolution.

Demographic transition does occur simultaneously with ongoing economic transition, education transition, health transition and reproductive health transition and all which affects the human development. There will be substantial improvement in human development and economic development if there is a association between these transitions. Historically in the western countries high level of urbanization and economic development accompanied by raising incomes have been associated with the nucleation of the families in terms of separate households (Fakhruddin 1991, p. 76).

Demographic status in Indian cities shows generally a bottom heavy youthfulness of the population and the family size and nature of family nuclear, joint, and extended are also having very important in the moulding of the specific urban way of life and culture. In Calicut city there is prominent culture of joint and extended family system especially in older residential areas like kuttichira, Mukhadar, which of Muslim population and Thiruvannur and Kallai area of caste Hindu population. Both of them have similarities in architecture of the houses regionally called '*Nalukettu*' and residing even more than five families and up to hundred family members in a single dwelling.

Table: 6.3 Family Status of Calicut City

S. No	*Variable*	*Factor Loading*
	HOUSING STRUCTURE	
1	R.C.C. Houses (per 100)	**0.4446**
2	R.B.C Houses (per 100)	**-0.4716**
3	Huts/ thatched Houses (per 100)	-0.2065
4	Owned houses (per 100)	-0.0364
5	Rented Houses (per 100)	0.0364
6	Houses with open space	0.0354
7	Houses without open space	-0.0314
8	Houses with 1-2 bedrooms (per 100)	-0.3273
9	Houses with 3-4 Bedrooms (per 100)	**0.5000**
10	Houses with above 4 bedrooms (per 100)	-0.1048
	AMENITIES AND FACILITIES	
11	Houses with sanitation Facilities	0.1968
12	Houses without Sanitation Facilities	-0.1924
13	Houses with Running Water	0.3334
14	Houses without running Water	-0.3326
15	Houses with telephone/ mobile	0.1323
16	Houses with no telephone/mobile	-0.1299
17	Houses With Electricity	0.1755
18	Houses without Electricity	-0.171
19	Household with Bank Account	0.3095
20	Household without Bank Account	-0.3072

Contd...

Table: 6.3 contd...

S. No	***Variable***	***Factor Loading***
21	Houses Subscribing News Paper (per 100)	0.2718
22	Houses with T.V	0.4039
23	Houses without T.V	-0.4041
24	Houses with Computer	0.3136
25	Houses without Computer	-0.3141
26	Houses with Car (per 100)	0.2393
	DEMOGRAPHY AND ETHINICITY	
27	Small household (per 100)	**0.8901**
28	Medium household (per 100)	0.0393
29	Large household (per 100)	**-0.9196**
30	Nuclear Family (per 100)	**0.9034**
31	Joint Family (per 100)	**-0.9034**
32	Hindu population (per 100)	**0.8075**
33	Muslim population (per 100)	**-0.8984**
34	Christian population(per 100)	**0.4218**
35	General Category (per 100)	**0.8537**
36	Backward Communities (per 100)	**-0.8666**
37	Work Participation Rate	**0.4782**
	EDUCATION AND INCOME	
38	Literacy Rate	0.3666
39	Family Head having High School Education	-0.3539
40	Family Head Having diploma/ graduation	0.3536
41	Low Income Household	-0.1192
42	Medium income Household	0.1723
43	High Income Household	-0.0525
	SECURITY FEELING AND MIGRATION	
44	Houses with High Security Feeling	0.2558
45	Houses with a feeling of Secure	-0.0015
46	Houses with Insecure Feeling	-0.2115
47	Inborn Households	-0.0263
48	Household Migrated within the city	0.0477
49	Household migrated from outside city	0.0001

Percentage of nuclear family and small households are the two highest loaded variables on this factor. In family structure the nuclear family have very important. The evolution from the extended family through joint family to the nuclear family is seen with the demographic transition of the modern communities.

Percentage of general category households has the higher loading on this factor. Normally the general category is composed of higher castes and affluent population and will have the economic and political command over the region. They will have easy access to the education, government and the higher facilities due to high economic back ground and support from the group which leads to a better quality of life as compared to the backward population. They always try to maintain their status and keep the minimum number of children, so that their demographic status will surpass.

Percentage of Hindu and Christian population is loaded positively. The Hindu population shares about 60 per cent of the total population in Calicut city. A sizable share of them is the descendents of the former rulers' *Zamorins* and their associates and the warriors of the *Zamorin* dynasty. So they possess a handsome hold in the location of settlement as well as in the quality of life. The Christian population are migrants from the surrounding area and districts. They are highly qualified and government officers and employed, and so their condition and status of living are comparatively high.

Work participation rate is defined as the percentage of total workers (main and marginal) to total population. It represents the qualification and capacity of the people. Work participation brings the better income and accessibility to the superior goods and services. The work participation rate is comparatively very less especially the female work participation rate in Calicut city. In Indian condition, traditional families are not in support to let the female work in public places or along with the males. But in some middle class families due to the financial stress and educational advancement, females began to engage in the jobs like teaching, clerical works, and some other professional jobs. Number of RCC houses is also loaded positively but marginally on this factor. As stated above holding the RCC houses represents the prosperity in income and high status of life in the society which partially reflect the high family status.

On contrary some variables are loading high but negatively on the dimension of the urban environment and the social wellbeing in Calicut city. The higher negative loading is reported by the percentage of the large household (-0.9197), followed by percentage of the joint family (-0.9035), percentage of the Muslim population (-0.8984), percentage of the Backward communities (-0.8667) and RBC houses (-0.4717)

The size of household has an effect on the affinity and hence tarnishes the quality of life and wellbeing of the city and society. The per capita income and time availability to children with parents are negatively correlated with the family size. It will affect the living status and mental ability of the new generation. The per capita indoor space is also related with the family size. So in general larger is the family size lesser is the quality of life explained by family factor.

Percentages of the joint family is also having high negative loading in this factor. It denotes joint family with all its advantages have some drawback with the social wellbeing at least in the case of Calicut city. Moreover in the marginalized population of the cities the system of joint family is not by choice, but is an obligation that deny the arrangement of a separate dwelling due to the economic stress. So the area having joint family in sometimes is a substitute of the low income families that is why in the social area analysis with weightage to housing and economic status, the area of the joint families should come under low categories devoid of their advantages.

In Indian condition low income and traditional families tend to extend family size so that pooled income can be raised to support the family differing from the western countries where the high level of urbanization and economic development led to nucleation of families in terms of separate households. This is further fostered to the kinship structure that is necessary to the achievement of political and material prestige. That is why the Indian traditional cities, in contrary to the western cities, have preserves the joint or extended family system. But in the case of Calicut city with its relation with the other world and the educational and cultural advancement shows more percentage of nuclear family in comparison to other same class Indian cities.

Muslim population and backward communities are exhibited marginally negative loading on this factor. These two variables have some relation in the context. Generally in India Muslims are categorized according to the ethnical character which originated based on the occupation of the ancestors. But this categorization is not affected the *Keralite* Muslims. There is no prominent ethnical categorization among them. As a minority all the Muslims in Kerala have given the other backward community (OBC) status by government. Apart this the majority of Muslim of Calicut city adopted the joint family tradition, especially those resides in the older residential areas. Percentage of RBC houses is also loaded negatively with a marginal importance can be relate to the less status of living indirectly to the family status.

The factor family status is emphasis that demography and ethnicity have an important role to determine the urban environment and social wellbeing along with a little importance to the housing structure in general and in Calicut city in particular. Positive loadings such as nuclear family, small household, percentage of general category, percentage of Hindu population, work participation rate, Christian population and RCC houses are favourable to the better urban environment and social wellbeing and the negative loading such as large household, percentage of the joint family, percentage of the Muslim population, percentage of the Backward communities, RBC houses are suppress the same.

Fig. 6.2 shows spatial pattern of the urban environment and the social wellbeing of the Calicut city expressed by the dimensions of the family status. Out of fifty-one eight wards shows very high factor score that means these wards are outstanding in family status as compared to other areas of the city. Sixteen scored high that means they are in good position in this status, six coming under the medium, fifteen wards are under low and the rest six are in very low category.

Very high region in the family status is more or less following the pattern registered by the factor material and housing condition and are also concentrated on the central parts of the city. This category comprising seven wards namely Kudilthodu, Cheveyur, Silver hills, Nellikkode, Kotooli south, Civil station, Kuthiravattom, on the centre and one on the southern parts namely Thiruvannur.

The area of very high scored in the family status is more or less following the high scored areas in the factor material and housing status. It signifies there is a close association between the family and better living with material and housing environment. Seven wards of central location in this factor shows these area is resided by government officers and highly educated population. It is trend in the educated people to restrict their offspring up to only one or two and live separate as nuclear family with the small household size. Most of this area is newly resided area that is why the traditional joint family system is very less. Work participation rate in this region also showing very high compared to other regions with both the couples are engaged in the employment. And the concentration of the negatively loaded variables like percentage of backward communities, percentage of the joint family and large households are rare in this region.

High scored wards in the factor family status are clustered mainly on the northern parts of the city and more or less adjacent to the very high scored region in the central parts. The wards coming under this category are Varaikal, Edakkat, East hill, West hill, Varaikal, Kuruvisseri, Malaparamba, Vengeri, Karaparamba, Nadakkavu, Chakkorathukulam, Cheverambalam, Kotooli north, Puthiyara, Calappuram, Azhchavattom, and Mankavu. Most of the region coming under this category is also the newly residential area and have the good condition in the family status. The work participation rate and the general category people are high in this region at the same time the joint family and the large sized families are comparatively less in this region. Peoples are opting nuclear family system and are small sized family.

Medium scored six wards have scattered location and the wards coming under this category are Paroppadi, Chevarambalam, Thiruthiyad, Eranchippalam, Meenchada and Kommeri. Low scored wards have a peripheral location covering entire eastern periphery. Poolakkadavu, Moozhikkal, Chelevoor, Mayanadu, Kovoor, Potamal, pokkunnu, Kinasseri, Panniyankara, Kallai, and Palayam are the wards coming under this category. And a coastal cluster comprising the wards New bazaar, Vellayil south, Vellayil north and Thoppayil are also coming in this group. This region is almost a replica of the medium category in the amenity and housing status which confirm the association between them. The area comes under this category due to the concentration of backward communities, joint families, and large sized households in some pockets of the region.

The very low scored wards are clustered as an elongated line along the coastal region in southern parts of the city. Very low scored wards in family status of the social wellbeing in Calicut city are, Big bazaar, Idiangara, Pallikkandi, Chakkumkadavu, Payyanakkal and Koyavalappu. This region comprising the old residential areas and the coastal belt exhibit the maximum concentration of the joint family, backward communities and large household families. Few pockets of this region has the population of highly qualified with high quality of life in all perspective but the overall condition of these wards obscuring by the vulnerable communities of the coastal region and those areas of slum like condition.

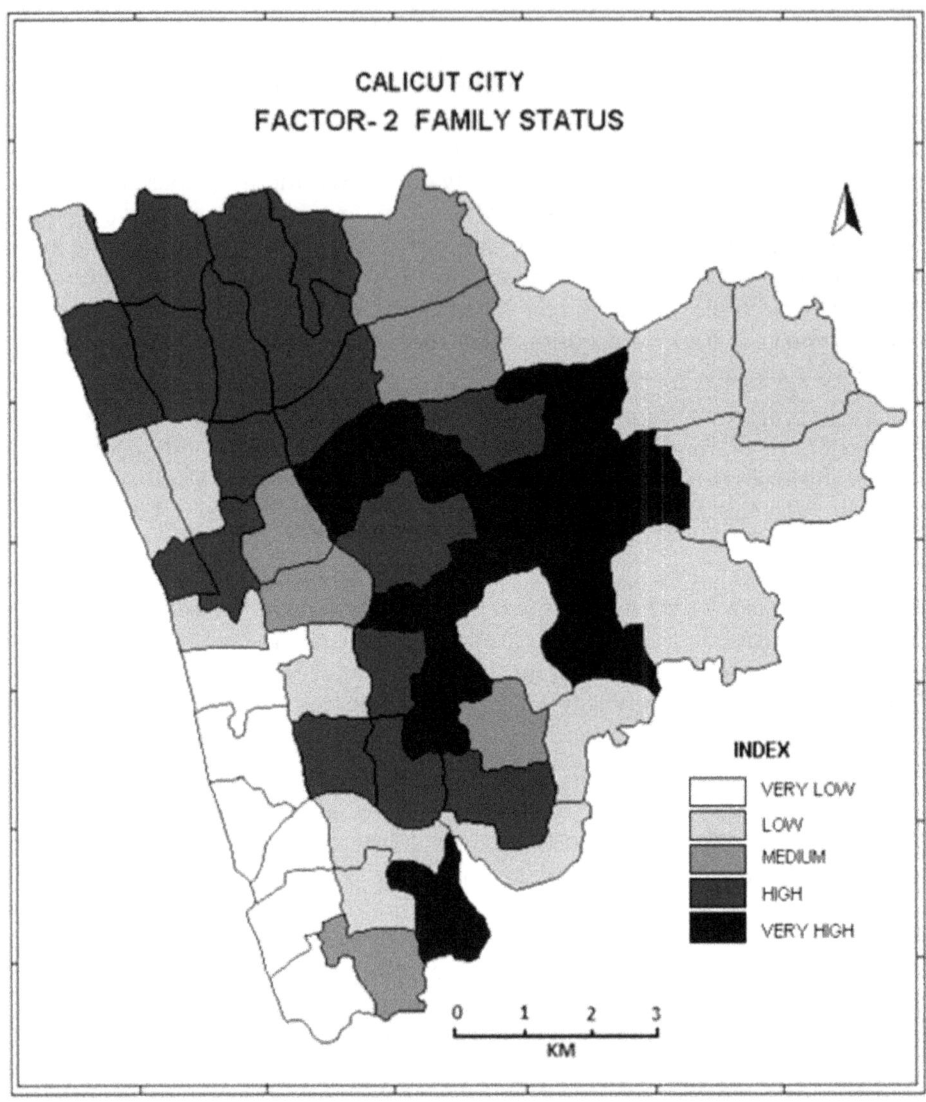

Figure: 6.2 Calicut City Factor-2 Family Status

The majority of the residents of the Calicut city belong to political, industrial, businessmen, and professional elites. They form the core of the modernized population in the city. It is therefore natural that the wards are characterized by small families; small proportion of children and a low fertility rate all emphasizing modernity of family.

6.3.3 Factor- 3 Education Status

This factor explains 17.22 per cent variance of the urban environment and social wellbeing in Calicut city. According to the factor score the education status is largely determined by fifteen indicators out of forty-nine. Four from the variable set education and income namely family head having high school education, Family head having diploma/graduation, low income household and high income household, and also four from the variable set of amenities and facilities namely houses subscribing news paper, houses with computer, houses without computer and houses with car. Four from housing structure that is houses with 3 or 4 bed rooms, houses with more than 4 bed rooms, RCC houses and RBC houses and two variable from the set security feeling and migration that is houses with highly secure feeling, houses with insecure feeling, and one from the demography and ethnicity factor that the percentage of Christian population.

The positive and negative association of the significantly loaded variables clearly shows that the factor 3 is the dimension of the un-education status. Those variables which support the education condition like high income, news paper subscription and presence of computer are scored negatively and those hamper the education like the low income, absence of computer and less per capita indoor space are loaded positively to this factor. But in order to maintain the positive nature of the factor head it is named as the education status in the place of un-education status.

Among the variable loading houses with 3-4 bed rooms scored highest (0.7862), followed by low income household (0.7197), houses without computer (0.7013), family head having high school education (0.6869), RBC houses (0.4653) and houses with insecure feeling (0.4250). 3-4 bed room houses denote the medium spaced houses in Calicut city and contribute maximum percentage of the total households of the city. It has a neutral effect on the education status but because of the high education status in the above 4 bed room houses that is among high income households have affected this factor.

Low income household scored second highest in this factor. It is very clear from the documented earlier studies that the income has direct relation to the education status of the individual, family and society. Generally the low economic or income status leads to low education status. Nowadays education especially higher education necessitates high capital investment specifically in private sector. In the case of professional education it is very difficult for the low income group to support their children for their studies even in the government sector. So the low income associates with the low status in the education factor.

Table 6.4 Education Status in Calicut City

S. No	*Variable*	*Factor Loading*
	HOUSING STRUCTURE	
1	R.C.C. Houses (per 100)	**-0.4480**
2	R.B.C Houses (per 100)	**0.4652**
3	Huts/ thatched Houses (per 100)	0.2668
4	Owned houses (per 100)	0.1711
5	Rented Houses (per 100)	-0.1711
6	Houses with open space	-0.1134

Contd....

Table: 6.4 Contd....

S. No	***Variable***	***Factor Loading***
7	Houses without open space	0.1199
8	Houses with 1-2 bedrooms (per 100)	0.3676
9	Houses with 3-4 Bedrooms (per 100)	**0.7861**
10	Houses with above 4 bedrooms (per 100)	**-0.8578**
	AMENITIES AND FACILITIES	
11	Houses with sanitation Facilities	-0.3004
12	Houses without Sanitation Facilities	0.3031
13	Houses with Running Water	-0.3877
14	Houses without running Water	0.3873
15	Houses with telephone/ mobile	-0.1150
16	Houses with no telephone/mobile	0.1178
17	Houses With Electricity	0.0211
18	Houses without Electricity	-0.0203
19	Household with Bank Account	-0.4068
20	Household without Bank Account	0.4063
21	Houses Subscribing News Paper (per 100)	**-0.5687**
22	Houses with T.V	-0.3747
23	Houses without T.V	0.3753
24	Houses with Computer	**-0.6983**
25	Houses without Computer	**0.7013**
26	Houses with Car (per 100)	**-0.6558**
	DEMOGRAPHY AND ETHINICITY	
27	Small household (per 100)	-0.1457
28	Medium household (per 100)	0.0958
29	Large household (per 100)	0.0512
30	Nuclear Family (per 100)	0.0168
31	Joint Family (per 100)	-0.0170
32	Hindu population (per 100)	0.0285
33	Muslim population (per 100)	0.1671
34	Christian population(per 100)	**-0.4420**
35	General Category (per 100)	-0.2805
36	Backward Communities (per 100)	0.2906
37	Work Participation Rate	-0.1811
	EDUCATION AND INCOME	
38	Literacy Rate	-0.2490
39	Family Head having High School Education	**0.6868**
40	Family Head Having diploma/ graduation	**-0.6863**

Contd....

Table: 6.4 Contd.....

S. No	Variable	Factor Loading
41	Low Income Household	**0.7196**
42	Medium income Household	-0.3868
43	High Income Household	**-0.8778**
	SECURITY FEELING AND MIGRATION	
44	Houses with High Security Feeling	**-0.5806**
45	Houses with a feeling of Secure	0.0321
46	Houses with Insecure Feeling	**0.4250**
47	Inborn Households	0.1418
48	Household Migrated within the city	-0.2868
49	Household migrated from outside city	0.0328

The variable houses without computer is also loaded high on this factor. Computer is the back bone of education in the era of education revolution. We can't assume even the primary education without computer. Denying this facility in the house is one of the symbols of the low income and low education status. Computer is the main technical component of the higher education, and without this the higher studies say to be incomplete.

Family head having only high school education is also loaded high and positively in this factor. The education of young family members and children are closely associated with the education and vision of the family head. If the family head is not well educated he/ she could not give good education to their offspring and could not help them in their studies. So the variable family head having only high school education denotes the low education status.

RBC house as stated in the above factors represent the middle income and low income groups in general and in the city in particular. It loaded positively in this factor which implies the low status of the education in those areas and households as the low income status inversely affect the education. Houses with insecure feeling are the expression of vulnerable environment and it too loaded positively to the factor education status. Education tends to the prosperity and the mentality to tolerate and understand the problems. If any place experiences insecure feeling it means the good share of people residing there have no education and culture. Moreover, the good education cannot be explored in such places where security is in casualty. So the positive association of the variable denotes the low education status in the area.

Some variables load high but negatively on this dimension. The highest negative loading is exhibited by high income household (-0.8779), followed by houses with more than 4 bedrooms (-0.8578), houses with computer (-0.6983), family head having diploma/ graduation (-0.6863) houses with car (-0.6559), household subscribing newspaper (-0.5688), houses with high security feeling (-0.5806), RCC houses (-0.4481) and percentage of Christian population (-0.4420).

Income as we stated above directly correlated with the education. Here high income household is loaded high negative that means the high scored region has less and low scored region have high status in this dimension. High income people have easy accessibility to the better primary and higher education which will lead them to high education status. Houses with more than 4 bed room also have very important association with the education status. It denotes the high income and prosperity and they will get calm and quiet environment to study as their per capita indoor space is generally very high as compared to the houses with 3-4 and houses with only 1-2 bedrooms.

The computer technology has a vital role and deep impact on education. Computer education forms a part of the school and college curricula, it is important for every individual to have the basic knowledge of computers. Students find it easier to refer to the Internet than searching for information in reference books. Physically distant locations have come close to each other only due to computer networking. Computers facilitate an efficient storage and effective presentation of information. The computer technology thus eases the process of learning. So those household having this facility in their own house denote the high education status, because the uneducated people cannot work easily with computer.

Family head having relatively high education is also loaded negatively in this factor of social wellbeing. On contrary to the family head with low educational qualification, the heads having high qualification will have good concern about the education of their younger ones and offspring. So in the members of family having well educated and visionary head, possess the high education status in comparison with others. They can help their children in their studies and can give good suggestions and advices in curricula and even in the problems of day to day life with a psychological touch.

Houses with car ownership which loaded negatively to this factor also elucidate relatively high income and the prosperity in economic condition. Those people with prosperity in economic condition, generally shows high education status than those having economic pressure, so the variable indirectly represent the positive relation with the education. The houses with secure feeling are also loaded highly that means the security condition favours the education. Children of those places experiencing security condition can go school and institutions without any fear, and the parents can send them school with confidence and feel free from any type of vulnerability.

Newspapers provide an abundance of interesting reading material on a daily basis. Newspaper-related activities enhance mastery of basic reading, writing and thinking skills and they are important in many other ways also. Reading newspapers everyday is a good habit for both students and adults for growth and enlightenment irrespective of the class or field of their life. It may consist of news reports, articles, photographs and advertisements. They have the main aims of informing current affairs, educating and entertaining the public. The youth are educated by it on social, political and economic issues and inform about the opinions and feelings of people in the world on various issues and about the new inventions, discoveries and developments in Science, Technology and many other fields. The households subscribing news paper denote the educational and social conciseness of the respective household. It is very necessary to update with the world affairs for the people as social being. The importance of the subscription and reading of news papers is only seen in the people who are educated. So it is an important factor in the determination of the education status of the household in particular, community and society in general.

RCC houses have also marginal negative loading in this dimension. It exert influence on high education status of the city that the RCC houses denotes the high standard of living in many aspects of life as indicated in the factor one i.e. material and housing status. And finally the percentage of Christian population is also has some contribution in this factor with marginal importance. The Christian community is comparatively new comers for the city and many of them migrated from other parts of the states after the appointment in the government sector and high profile private sector. Apart from this generally Christian population is far better in education, employment and hence in socio economic condition than other communities in Kerala in general and in Calicut city in particular.

Considering the magnitude of the loadings of significant variables in this factor, an environment characterized by housing space, facilities, educated family head and the favourable income is significant for the healthy urban environment and the social wellbeing in the cities as well as surrounding villages.

The positive association of the variable like medium indoor spaced houses, low income household, houses without computer, and family head having high school education, RBC houses, houses with insecure feeling which are the indicative of the low socioeconomic and education status and the negative association of the variable like high income household, followed by houses with more than 4 bedrooms, houses with computer, family head having diploma/ graduation, houses with car, household subscribing news paper, the houses with high security feeling, RCC houses, and percentage of Christian population which represent the high socio economic and education status indicate that factor 3 can be identified as the dimension of the un-education status.

Fig 6.3 shows the spatial pattern of the social in Calicut city expressed by the factor education status. As this factor have positive association with those variables leading to poor education status and negative association with the variables support high education, the higher the score in the component represent the lower the education status and vice versa. Only three wards out of fifty-one have very low factor score that is they show the outstanding condition in the education and higher facilities as compared to other wards of the city. Fifteen wards records the low factor loading that exhibit the good condition, another fourteen ward shows medium score expressing the medium condition, twelve wards coming under the title high and the rest seven ward under very high factor score denotes the below average and poor condition respectively.

The very high scored or poor standard wards in the education status is the coastal stretch embrace the wards Koyavalappu, Payyanakkal, Chakkumkadavu, Pallikkandi, Vellayil south, Vellayil north and Thoppayil. These wards are the most vulnerable region in the city in almost all the factors. The region is inhabited by predominantly by fishermen and labours. They are not educated to teach their younger ones properly and the low income and the poor socio-economic condition lead to low education status in this region. Percentage of newspaper subscription and the presence of computer in the household are lowest in this region.

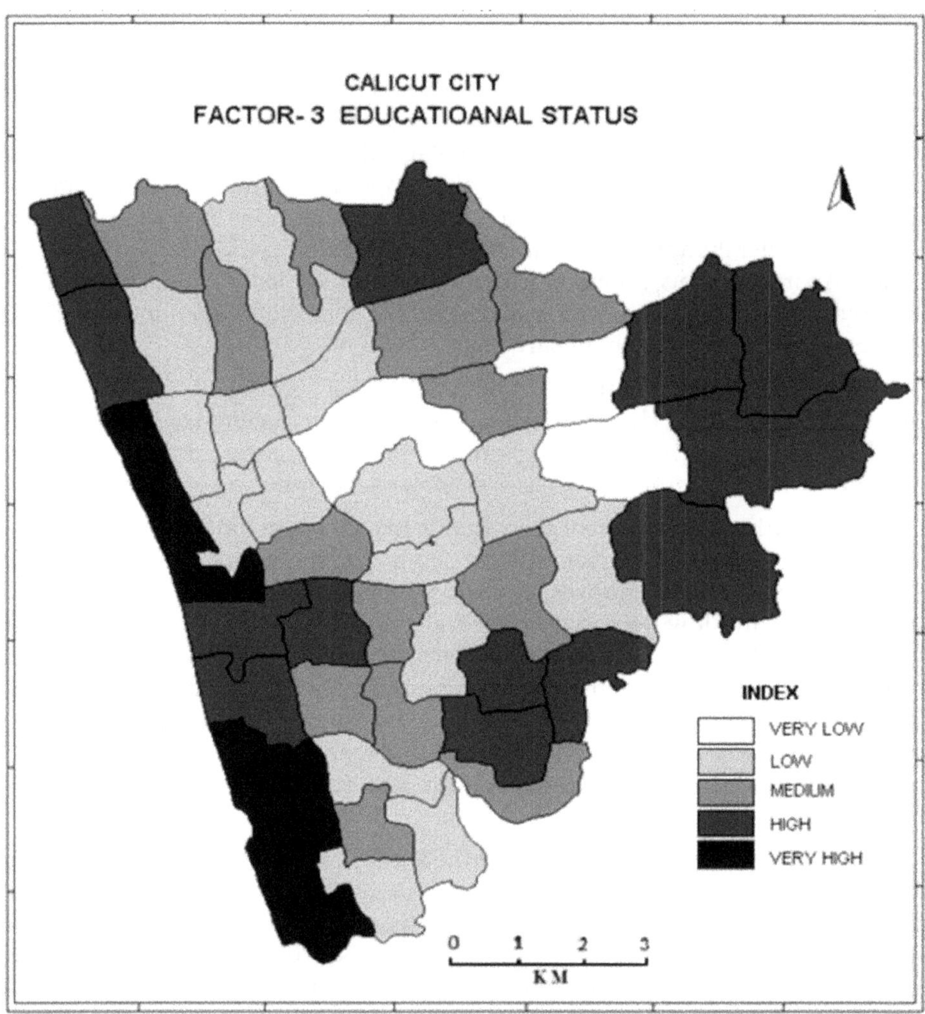

Figure: 6.3 Calicut City Factor-3 Education Status

Likewise medium scored wards are located as two separate clusters, one in the northern part of the city comprising the wards Edakkat, East hill, West hill, Vengeri, Paroppadi, Chevarambalam, Poolakkadavu. And the southern cluster includes the wards Kinasseri, Panniyankara, Chalappuram, Azhchavattom, Puthiyara, Potamal, and Tiruthiyad. This region composed mostly of middle income household and have good share of RBC houses both of them adversely scored to the education factor. The education of the family head and the percentage of household having computer and news paper subscription are comparatively less which suppress the education status in this region.

Low scored wards in this respect are clustered almost adjacent to the very low region in the central part and another small cluster in the southern part of the city. Central cluster comprising twelve wards- Kuruvisseri, Malaparamba, Karaparamba, Nadakkavu, Eranchippalam, Varaikal, Kotooli north, Kotooli south, Kuthiravattom, Nellikkode, Kudilthodu and Chakkorathkulam and the southern cluster includes only three wards- Thiruvannur, Kallai, and Meenchanda. This area is showing high status by a reason of the high percentage in news paper subscription, presence of computer in the houses and also the presence of high income population. These areas have several educational institutions especially in the secondary and higher secondary schools which gives a basement to the modern and high education. Education among the seniors and family heads are furthermore showing high rate in this region.

Very low scored region in the un-education status, that is the highest ranked region in education lying on the geometrical centre of the city comprising only three wards- Chevayur, silver hills and Civil station. These wards are the posh residential area of the city in almost all respects which we considered in this analysis. This area is mostly resided by high and medium income groups and the percentage of low income households which deteriorate the education status is very less in this region. Moreover this area especially Silver Hills and Chevayoor are the education hub of the city in secondary and higher secondary education. As we stated above a good share of the households have government employs and so their offspring are also educated.

6.3.4 Factor- 4 Security Status

Factor Four which explains 10.41 percentages of variables in the urban environment and social wellbeing of the Calicut city can be broadly identified as the dimension of the security status. The security status in the city is determined largely by seven variables out of forty-nine, four of them related to the variable set of housing structure namely percentage of owned house, percentage of rented houses, percentage of houses with open space and percentage of houses without open space and three related to the variable set security feeling and migration include the households with a feeling of 'secure', household with a feeling of 'in secure', and the household migrated from outside the city.

Among the variable loading percentage of owned houses (0.8902) loads highest followed by houses with 'secure' feeling (0.7822) and percentage of houses with open space (0.7084). Housing insecurity is an issue that affects the population and one of the most important aspects to survival of family and community. It reflects major trends in income distribution, culture and aesthetic factor of the region. Generally people pay out a noble per cent or more of their income on housing for a better and safe shelter.

Table 6.5 Security Status of Calicut City

S. No.	*Variables*	*Factor Loading*
	HOUSING STRUCTURE	
1	R.C.C. Houses (per 100)	0.2975
2	R.B.C Houses (per 100)	-0.2979
3	Huts/ thatched Houses (per 100)	-0.2118
4	Owned houses (per 100)	**0.8901**
5	Rented Houses (per 100)	**-0.8901**
6	Houses with open space	**0.7083**
7	Houses without open space	**-0.7101**
8	Houses with 1-2 bedrooms (per 100)	-0.2591
9	Houses with 3-4 Bedrooms (per 100)	0.1885
10	Houses with above 4 bedrooms (per 100)	0.0613
	AMENITIES AND FACILITIES	
11	Houses with sanitation Facilities	0.2184
12	Houses without Sanitation Facilities	-0.2197
13	Houses with Running Water	0.1936
14	Houses without running Water	-0.1920
15	Houses with telephone/ mobile	0.1017
16	Houses with no telephone/mobile	-0.1040
17	Houses With Electricity	0.1236
18	Houses without Electricity	0.1246
19	Household with Bank Account	-0.1343
20	Household without Bank Account	0.1346
21	Houses Subscribing News Paper (per 100)	0.1961
22	Houses with T.V	0.1051
23	Houses without T.V	-0.1050
24	Houses with Computer	0.0010
25	Houses without Computer	-0.0023
26	Houses with Car (per 100)	-0.0532
	DEMOGRAPHY AND ETHINICITY	
27	Small houshold (per 100)	-0.0835
28	Medium household (per 100)	0.3127
29	Large household (per 100)	-0.1824
30	Nuclear Family (per 100)	0.1159
31	Joint Family (per 100)	-0.1165
32	Hindu population (per 100)	-0.0096
33	Muslim population (per 100)	0.0081
34	Christian population(per 100)	0.0234

Contd...

Table: 6.5 Contd....

S. No.	*Variables*	*Factor Loading*
35	General Category (per 100)	-0.0604
36	Backward Communities (per 100)	0.0721
37	Work Participation Rate	-0.2911
	EDUCATION AND INCOME	
38	Literacy Rate	-0.0043
39	Family Head having High School Education	-0.1750
40	Family Head Having diploma/ graduation	0.1750
41	Low Income Household	-0.0131
42	Medium income Household	0.0835
43	High Income Household	-0.1175
	SECURITY FEELING AND MIGRATION	
44	Houses with High Security Feeling	-0.1110
45	Houses with a feeling of Secure	**0.7821**
46	Houses with Insecure Feeling	**-0.7351**
47	Inborn Households	0.3047
48	Household Migrated within the city	0.1134
49	Household migrated from outside city	**-0.4961**

Percentage of houses with secure feeling is loaded highly and positively on this factor. Security feeling is mostly of perception and experience. In some region of the Calicut city there are residential associations and self help groups to care about the street and neighbourhood. Urban environment and social wellbeing are closely associated with the security feeling because people like to live in calm and quiet atmosphere. Percentage of houses with open space is also loaded positively. It designate that the area having much open space in their residences have good security feeling. This relation is nothing but the open space are commonly available at the newly settled residential areas were security feeling is good, and in congested areas people may not know even their immediate neighbours. In the outskirts of the city people are well familiar each other and no need to worry about their houses and properties even in their absence because generally the neighbours will be cautious about it.

Four variables loaded high but negatively on this dimension. The highest negative loading is registered by percentage of rented houses (-0.8902), followed by percentage of without open space (-0.7101), houses with 'insecure' feeling (-0.7351), and percentage of household migrated from outside the city (-0.4961). Percentage of rented houses in contrary to the owned houses will insert bad effect on the security feeling and the wellbeing. The community involvement and attachment will be less in the rented households as a result of non permanency and non surety of their living place. So making a good relation with the neighbouring people is not so easy for them. Moreover for the owned houses people they feel the neighbourhood and the streets as their own, but for the rented people it will feel as non belongingness. It is also related with the housing security i.e. for those having own houses are secure in housing and those don't have own houses are insecure in housing.

Percentage of houses without open space is also loaded high but negatively. It means that the areas of less or without open space are congested areas and devoid of personal and family contact with each other and they can't familiarize with the face of who passes through their area. Houses with insecure feeling are also having a direct impact on the security feeling and hence the overall urban environment and wellbeing. The insecurity feeling may be caused by the high crime rate, communal sensitivity, vulnerability to natural and health hazards and the weakness of the houses and housing materials. Percentage of the households migrated from outside the city is loaded negatively but marginally to this factor. It is very common that the newly migrated people will take time to adjust and familiarize with the new urban environment and people. So they will be in dilemma with the environment and inhabitants. They will be stranger for the local inhabitants at least for some times, that may may guide to a insecurity feeling. For the newly migrated population the security feeling will be lesser than the permanent occupant even though they select the places where they can live serenely.

The wards scoring high in this factor shows the high housing standard and security feeling which is a legacy of the strength of building materials, the cooperation and support of the neighbouring people and the immediate environment and the ownership of the dwelling. The wards scoring low in this factor symbolize the insecurity feeling by the weak building materials strangeness to the neighbouring people and immediate environment and the congestion on the dwelling area.

Fig 6.4 shows the spatial pattern of the urban environment and social wellbeing in Calicut city articulated by the dimension of the housing and security status. Out of fifty-one wards five shows the very high factor score that is the hosing space is excellent and the insecurity feeling is very less in these area, seventeen wards showing high factor score that the housing space and the security condition is well in this region, eighteen wards scores medium that denotes the situation in this respect is average in that area, eight wards records low and only three very low that means the congested hosing condition and the fear of crime and theft is prevailing in that area.

The very high region in the housing and security status is concentrated on north eastern peripheral region of the city including five wards- Mayanadu, Chelvoor, Mozhikkal, Poolakadavu and Kannadikkal. In this region land value is comparatively less and the open space and housing space is abundant. The region is totally residential area so the commuting of outsiders and strangers are very less, so there is no casualty of the security condition. The interaction between the households is prominent since a village like condition is prevailing in these areas, and the situation of robbery and theft is very less. Moreover the percentage of rented houses is very less in this region which reduces the security threats of the region. And the households migrated from outside the city is also recorded very less that is a variable which affects the security feeling negatively.

The high scored region in this factor is stretched over the peripheral region except the western coastal region comprising fourteen wards namely New bazaar, Varaikal, Edakkat, Kuruvisseri, Karaparamba, Vengeri, Paroppadi, Chevarambalam,

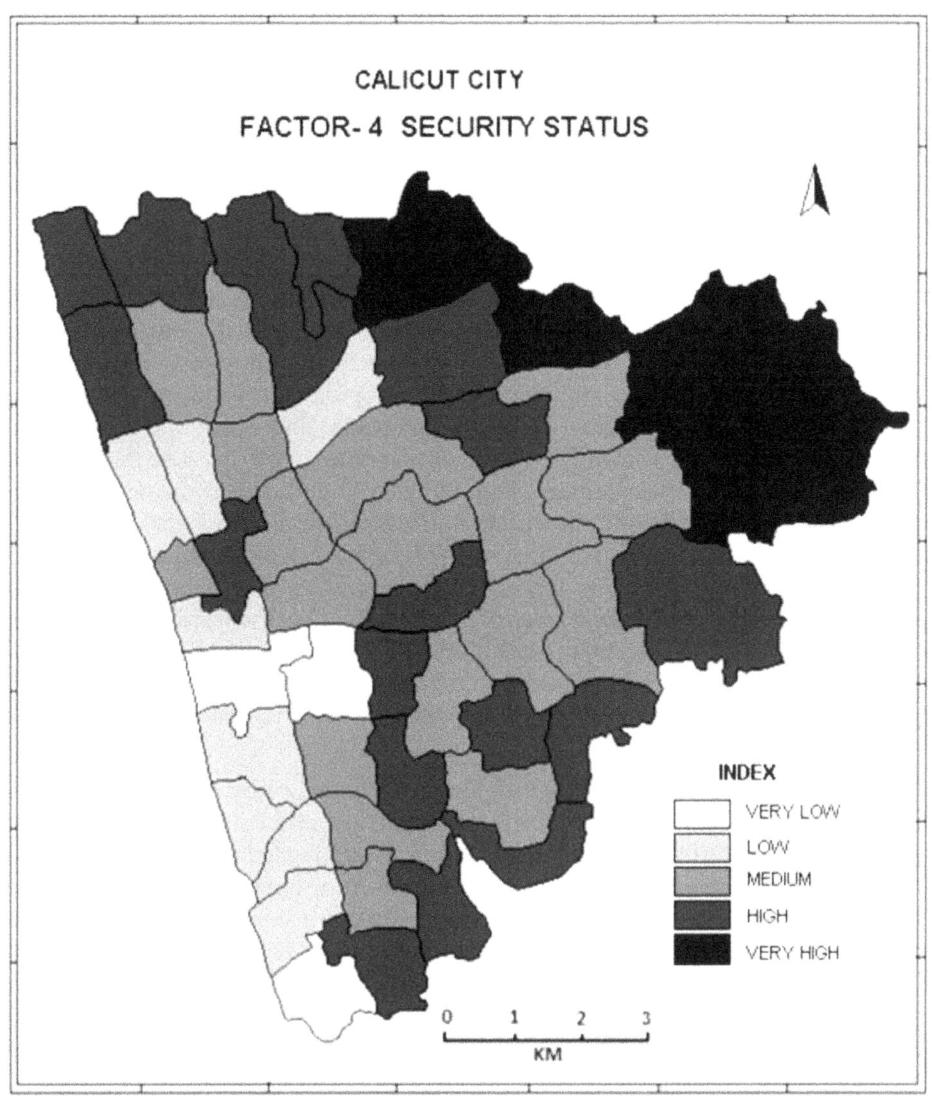

Figure: 6.4 Calicut City Factor-4 Security Status

Kovoor, Kommeri, Pokkunnu, Kinasseri, Thiruvannur, Meenchanda and the inner cluster of three wards that is Kotooli south, Puthiyara and Azhchavatom. These areas are not so behind the very high scored region. These peripheral regions also have good relation among people and are exclusively residential areas. Most the houses of these have housewives who remain almost twenty four hours in the houses.

Another of the important reasons to score in this category is the presence of comparatively large number of rental houses due to the locational advantage to several institutions, government offices and companies.

The medium scored wards in the factor of housing and security status is lying in the middle parts of the city stretching over eighteen wards namely, West hill, East hill, Chakkorathukulam, Nadakkavu, Eranchippalam, Thiruthiyad, Civil station, Kotooli north, Kudilthodu, Chevayur, Silver hills, Nellikode, Pottamal, Kuthiravatom, Mankavu, Chalappuram, Kallai and Panniyankara. The housing space in most of these wards of this region is good but due to the low loadings in the security condition these wards came under the medium category. The area is oriented to business enterprises and small scale industries so that the commuting of people is often in this area which leads to the security feeling under stress.

Low scored wards in housing and security status is spread along the coastal area covering most of the coastal wards namely, Payyanakkal, Chakkumkadavu, Pallikkandi, Idiangara, Vellayil south, Velleyil north and Thoppayil and one inner ward Malaparamba. The coastal belt is thickly populated area in the city, and the housing condition is very poor with the low social and economic condition of fishermen settled there. Security feeling is also very less there because they have no facility to lock safely their valuable belongings and also due to the occurrence of crime and quarrel between the inhabitants. The housing condition and tenor ship is also a big problem in this region. The socio economic condition leads to give some parts of their houses for rent even though they are in stress due to the below average per capita indoor space. In some areas of coastal settlement there is a vulnerability of communal violence among the people. Moreover coastal villages have vulnerability of sweeping their villages by the tides in the high rainy months of monsoon. In the case of Malaparamba the presence of NGO cottages which consist about 300 rented households deteriorating the score.

Only three wards recorded the very low score, namely, Koyavalappu, Big bazaar and Palayam, in which Koyavalappu is a coastal ward, and Palayam and big bazaar are the market place from where people largely migrating towards the outer region. These two wards coming under the 24 hour commuting area, so the security is most vulnerable and the housing is also very poor with old residences, no one willing to newly settle in this area and the land value is very high due to the market location. The people present there is also wish to move outwards of the city where a good open space and environment is available and the area becoming fully commercial with establishment of new markets and storage places. In the case of the ward Koya Valappu it is the most vulnerable coastal settlement having vulnerability in the security lying in the south western corner of the city and is near to 'Maradu' a settlement were communal violence is occurred in 2003.

6.3.5 Factor- 5 Migration Status

Factor 5 which explain 08.44 per cent variance of the social wellbeing in Calicut city can be identified as the dimension of the migration status. According to the factor score the migration status is largely determined by only four indicators out of forty-nine. Three from the variable set of security feeling and migration namely inborn household, household migrated within the city and household migrated from outside the city and one from the variable set of demography and ethnicity namely medium sized household.

Table 6.6 Migration Status of Calicut City

S. No.	Variables	Factor Loading
	HOUSING STRUCTURE	
1	R.C.C. Houses (per 100)	-0.1704
2	R.B.C Houses (per 100)	0.1617
3	Huts/ thatched Houses (per 100)	0.1265
4	Owned houses (per 100)	0.2135
5	Rented Houses (per 100)	-0.2135
6	Houses with open space	-0.2062
7	Houses without open space	0.2034
8	Houses with 1-2 bedrooms (per 100)	0.2078
9	Houses with 3-4 Bedrooms (per 100)	-0.0057
10	Houses with above 4 bedrooms (per 100)	-0.1494
	AMENITIES AND FACILITIES	
11	Houses with sanitation Facilities	-0.1912
12	Houses without Sanitation Facilities	0.1906
13	Houses with Running Water	-0.1666
14	Houses without running Water	0.1667
15	Houses with telephone/ mobile	-0.1608
16	Houses with no telephone/mobile	0.1607
17	Houses With Electricity	-0.0673
18	Houses without Electricity	0.0699
19	Household with Bank Account	0.1030
20	Household without Bank Account	-0.1027
21	Houses Subscribing News Paper (per 100)	-0.1030
22	Houses with T.V	0.0354
23	Houses without T.V	-0.0373
24	Houses with Computer	-0.3518
25	Houses without Computer	0.3528
26	Houses with Car (per 100)	-0.1652
	DEMOGRAPHY AND ETHINICITY	
27	Small household (per 100)	0.2535
28	Medium household (per 100)	**-0.7028**
29	Large household (per 100)	0.3371
30	Nuclear Family (per 100)	-0.2382

Contd...

Table: 6.6 Contd....

S. No.	*Variables*	*Factor Loading*
31	Joint Family (per 100)	0.2396
32	Hindu population (per 100)	0.1050
33	Muslim population (per 100)	0.0941
34	Christian population(per 100)	-0.3770
35	General Category (per 100)	0.0931
36	Backward Communities (per 100)	-0.1037
37	Work Participation Rate	-0.2202
	EDUCATION AND INCOME	
38	Literacy Rate	-0.1859
39	Family Head having High School Education	0.2212
40	Family Head Having diploma/ graduation	-0.2222
41	Low Income Household	0.1809
42	Medium income Household	-0.3996
43	High Income Household	0.1173
	SECURITY FEELING AND MIGRATION	
44	Houses with High Security Feeling	0.2397
45	Houses with a feeling of Secure	-0.3137
46	Houses with Insecure Feeling	0.1368
47	Inborn Households	**0.8933**
48	Household Migrated within the city	**-0.8467**
49	Household migrated from outside city	**-0.5293**

Among the variable only the inborn household (0.8933) loaded positively. It means the inborn household favours the urban environment and social wellbeing as local residents are quiet aware and adopted with the local environment and people that should make a harmony between them. Moreover even the modern era migration is a serious problem and yet to be tackled in proper way. So those areas where inborn population is very small shows either the people from this area migrated to other places due to the uncomfortable social or physical environment or both. Similarly it denotes the large number of people settled here from due to better quality environment with higher facilities and job opportunities.

Three variables are loaded negatively on this factor namely household migrated within the city (-0.8467), medium sized household (-0.7028) and household migrated from outside the city (-0.5293). Generally the region where the new migration is taken place will be well maintained and established residential areas or vacant lands where people began to colonize. It represents the good and calm environment is prevailing there.

Likewise the medium sized household is loaded negatively on this factor of urban environment and social wellbeing. Medium sized families always have their advantages and will be psychologically stronger than both the small and large sized families. So the region having high concentration of medium sized household is well-off in social environment. Percentage of household migrated from outside the city in addition loaded negatively but only marginally to this factor. This factor can have two way direction i.e. positive and negative effects in social environment. Positive element is that the incoming population along with high socioeconomic condition will choose the region where a good environment is prevailing whereas the negative elements is that the social harmony will be lesser as the people are from different social environment. In addition if the immigrants are low socioeconomic class they shall prefer the city centre with unhealthy condition for easy access to their work place.

It is also observed that people moves or wish to move from only such area were the environment and wellbeing is relatively poor. So those place from where people moves to other areas are either congested or less favourable in the sense of environment and wellbeing, on the other hand where the inborn population is less and immigrant population is high the region may be a newly-resided area where the satisfactory condition of social and physical phenomenon is existing. So the migration factor can be explained both optimistically and undesirably.

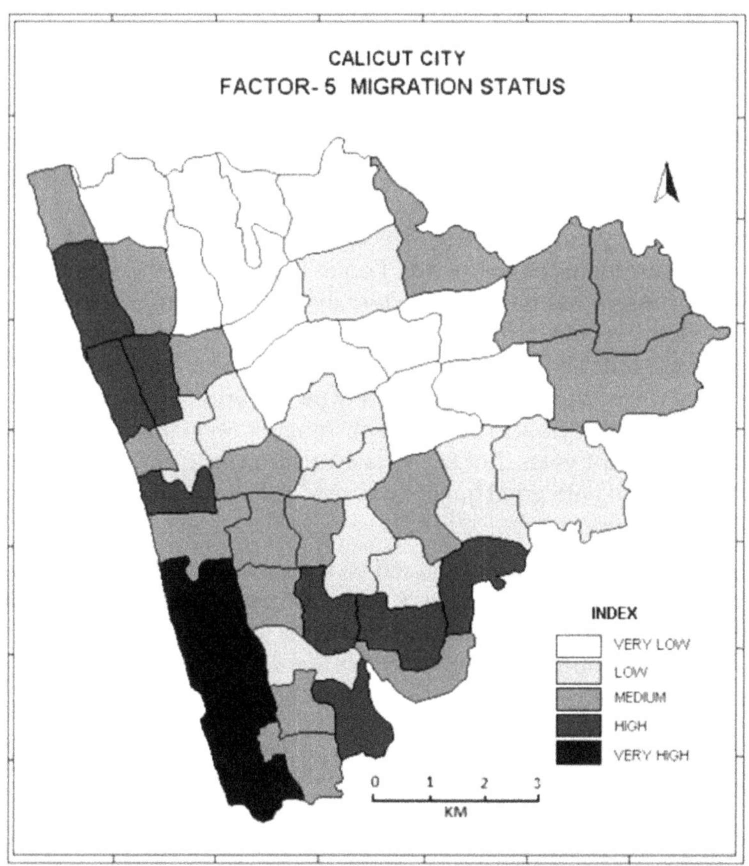

Figure: 6.5 Calicut City Factor-6 Migration Status

Fig 6.5 shows the spatial pattern of the social wellbeing in Calicut city expressed by the factor migration status. Only five wards show very high factor score in the migration status that means these wards experience less liveability as recorded by this factor in comparison to other wards of the city. Eight wards records the high factor loading that means the condition is better than of very high scored area but yet people are wish to move from this area due to several reasons, another seventeen ward shows medium score expressing the medium condition, ten wards coming under the title low and the rest eleven ward under very low factor score denotes there is only a little outward and inward movements of people from these areas.

Very high scored region in migration status lying on the south western coastal line of the city comprising five wards namely Koyavalappu, Payyanakkal, Chakkumkadavu, Pallikkandi and Idiangara. If we compare the other factors of the analysis these areas recorded least scores in the material and housing, education, family and security feeling. So we can understand the area is not so good for living and the high negative association implies that the in and outward movement of population is very less in this region and can be say it is a stable settlement.

High scored wards in this respect are clustered northern half of the coastal line covering Varaikal, Thoppatil, Vellayil north and Vellayil south, and a small cluster on the south-eastern peripheral part of the city including Thiruvannur, Pokkunnu, Mankavu and Azhchavattom. These areas are also having somewhat stable condition and least migration is taking place in this region. These wards are comparatively well-off parts of the older settlement.

Medium scored wards are also located as two separate clusters, one in the western part of the city lying almost parallel to the coastal line comprising the wards New bazaar, East hill, Chakkorathukulam, Nadakkavu, Big bazaar, Tiruthiyad, Chalappuram, Palayam, Puthiyara, Menchanda and Panniyankara, and eastern peripheral cluster includes the wards Poolakkadavu, Mozhikkal, Chelavur and Mayanadu. Kinasseri and Pottamal are two unclustered wards in this respect. This wards shows there is both the immigration and migration taking place. This region is not so congested but the land value is very high. In the western cluster the people going outward due to the congestion but new people are coming due to the nearness to the their offices markets and other establishments. So in this area the apartments are piling up in recent years. But in the case of eastern cluster people prefer it as a newly settled region with good quality physical and better social environment with village like condition.

The low scored wards are clustered on the south- eastern region covering the wards Kovoor, Nellikode, Kommeri, Kuthiravatom, Kotooli south, Kotooli north, Eranchippalam and Karaparamba and Paropadi laying un-clustered in northern parts and Kallai in southern parts. In these wards new settlements are amassing owing to the good physical environment with open spaces and greenery and the accessibility and nearness to the city centre.

The very low scored wards in this respect shows a clear clustering in the northern and north central parts of the city comprising eleven wards Edakkat, East hill, Kuruvisseri Vengeri, Kannadikkal, Malaparamba, Civil station, Cheverambalam,

Chevayur, Silver hills and Kudilthode. These wards are the posh area of Calicut city as considering the residential establishments and the inhabitant's wellbeing. So the people prefer this area to settle down. And the presence of the good facility to health and higher education magnetize the high and medium income population to here.

6.3.6 Overall Wellbeing in Calicut City

In order to get an overall image of the urban environment and social wellbeing represented by all the five factors in Calicut city a composite factor is obtained and represented on a map. For this purpose the standard component scores for each observation or wards are added and divided with the total number of factors. One can easily understand that which parts or wards of the city is most habitable in the sense of overall quality of life in both the physical and social environment and which places are the least comfortable for living in comparison to the other region with the help of overall wellbeing.

All standardized component scores are taken in to consideration in crude, and the factor 3 i.e. the education status and factor 5 migration status are used in a different way that needs more explanation. The crude data of education factor with its positive and negative associations can be identified as the non education status out of education status. If we use the scores as it is, it will alter the actual result as this show negative standard and all other are positive. So without any alteration in the magnitude, the sign of the standard component scores of the education status is changed from positive to negative and vice versa. The fifth factor that is migration status given only half weight as it can affect the area both favourably and unfavourably (Appendix. 1). So we can express the composite factor as [F1+F2+ (- F3) +F4+ (F5/2)]/ 5 and the result is plotted in the map. The composite factor will show the overall urban environment and social wellbeing explained by forty nine variables and five factors

Fig 6.6 shows the composite factor or the overall urban environment and social wellbeing recorded by five factors and forty nine variables. The spatial pattern of the composite factor divides the city in to five distinct regions i.e. the inner core of highest, the central parts of the high, eastern periphery of medium the northern coastline of low and the southern coastline of the very low overall urban environment and social wellbeing.

The very high region comprising the wards Civil Station, Kotooli South, Kudilthodu, Chevayur, Silver Hills and Nellikkode clustered in the central parts, and two isolated wards Westhill and Thiruvannur in north and southern edge respectively. Very high index region shows the best region to reside and is comparatively newly settled region developed as the residential area by both the municipal corporation and the residents themselves. These areas resided by the medium and high income groups and have a good security feeling among the residents from crime and communal violence. Most of the people are well educated and have a good understanding of the society matters. Almost all the variables which boost the five factors of the analysis i.e. material and housing, education, family, security status and the migration are scored highly on this region.

High composite index region have a north south stretch though the central parts of the city having twenty wards namely Varaikal, Edakkatt, East hill, Kuruvisseri Malaparamba, Vengeri, Paropadi, Chevarambalam, Kotooli North, Thiruthiyad, Eranchippalam, Karaparamba, Nadakkavu, Chakkorath kulam, Kuthiravattom, Puthiyara, Kallai, Chalappuram, Azhchavattom and Meenchanda. High loaded category region are also performing good condition in the urban environment and the social wellbeing represented by the five factors. But the reason for the position behind the very high region is that some variables are loading negatively or positively but only moderately to this region. We cannot say specific variable(s) is loading negatively or moderate positively to all region, but every ward(s) may have exclusive demerits like the congestion, presence of slum sites, old and outdated residential establishments, and low security feeling. If anyone ready to scarifies such minor demerits, this area is also better to reside in.

Medium scored wards have organized in the peripheral parts in eastern and south eastern region of the city and comprising twelve wards namely Panniyankara, Kinasseri, Potamal, Komeri, Pokkunnu, Mankavu, Poolakkadavu, Moozhikkal, Chelavoor, Mayanadu and Kovoor and New Bazar as dispersed in the north western corner of the city. Medium scored region have some what a balanced loading in factors. The most important thing is that there is no casualty of the indicators like security feeling, open spaces and inborn population, but there is no facility of higher education and higher health care and they depend on the city centre for their needs. The presence of the low income households, RBC houses, comparatively low education status of family heads are led the region in to the average condition. The main distinctiveness is that there is no harmful condition prevailing but only some inadequacy of some higher facilities. If the government is willing to provide such facilities in this area, it can develop a high profile residential area.

Low composite index region have completely coastal stretch except Palayam which is market place and other wards are Pyyanakkal, Idiangara, Big bazaar, Velleyil south and Vellayil north. Very low index region is also have coastal extend including the wards Koya valappu, Chakkumkadavu and Pallikkandi. Almost all the factors and all variables which support the urban environment and social wellbeing are loaded negatively in this region. These areas have the highest house density in the city and most of the residing people are labour class and fishermen. Some slum sites of Calicut city like Bangladesh colony, Chamundi Valappu, Kodi, Ninan Valappu are present in this region. These wards are composed of older residential areas and have very congested with low and medium income RBC houses. RCC houses, high income households, business establishments, open spaces and the educational institutions are very less in this region. People of this area are not so educated and most of them are labours in primary and secondary sectors. The case of wards Palayam and Big bazaar is something different from that of others, they became fully established market places and 24 hour commuting area. The security feeling and serenity are in casualty with the non-permanent residents and old and unmaintained residential establishments. Koya valappu, Chakkumkadavu and Pallikkandi are the most vulnerable area in this region.

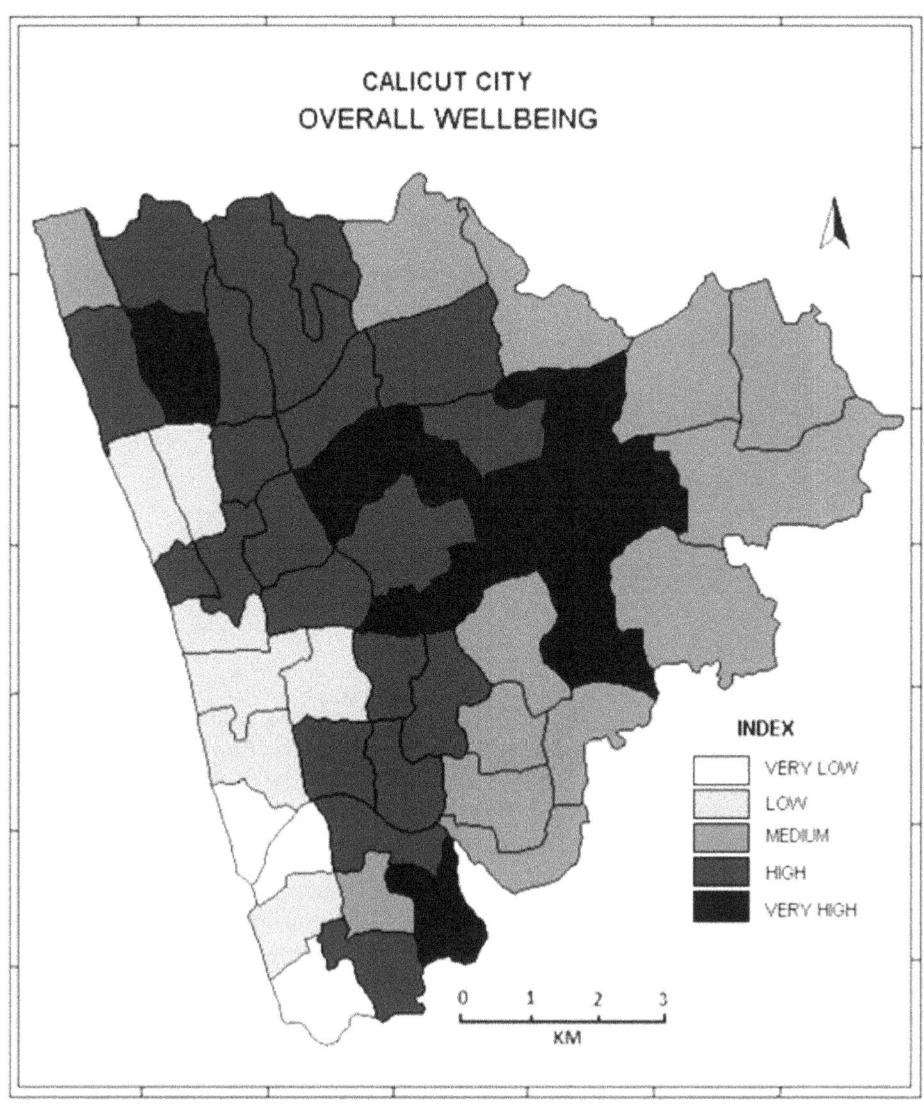

Figure: 6.6 Calicut City Overall Wellbeing

CONCLUSIONS

The present study of the urban environment and social wellbeing in Calicut city despite its limitations have implications in bring forth certain important features of the city space and the general urban environment and social wellbeing in the context of developing countries. We can draw some specific findings, conclusions and broad generalizations from the observations and analysis.

The historical background of the origin and development has influenced the specific culture and spatial organization of the residential and commercial land use of Calicut city. Historically the city has been the commercial and cultural capital of the Malabar region and the communal harmony and social relations are preserved from the beginning to the present. The historical background also influenced on the layout and the residential pattern of the city.

The old city region have two segregated residential areas one of which is Muslim dominated region of Kuttichira and surrounding area in the centre portion of the coastal low land area and other is of Hindu domination Thiruvannur and adjoining areas on the south eastern corner of the city. The older settlements of these two sections resembles in their architecture, and even the family organization practicing joint family system. The cast Hindu population more advancing towards nuclear family set up but the Muslim in this region is almost consistence practicing the joint family system and the matriarchal system of succession. It is very interesting that the oldest mosque (Mishkal mosque in Kuttichira) and temple (Tali Temple) also resembling in their architecture and moreover both of them have large ponds in their vicinity.

The modern urbanization is somewhat complex in its features. Most of the urbanization followed the main roads connecting the city to outer region. The settlements of the Civil Station, Malaparamba and Silver Hills are originated in the administrative purpose and have maximum number of government employs and professionals. The recent extension of the city is towards the peripheral region in eastern and north eastern side and along the coastal line towards both northern and southern side. Overall Calicut city is not a congested city. Inter house spacing and open spaces are very much in comparison to other equally populated Indian cities because of the unique *keralite's* housing culture that a house must carry open spaces in all sides. The modern residential pattern under the neo socio-political and economic organization, following the economic, employment and social status features rather than the cast and creed.

The coastal area, a considerable share of which represent the old city nothing to say modernization and this part is comparatively congested and resided by comparatively poor sections of the population including fishermen. The under developed and deteriorated social and physical environment is prominent in this region. This state of under development is partially because of the congestion along with the ignorance of the municipal authorities and government. Even the primary human needs like sanitation and drinking water facilities are not so satisfactory in this region. Moreover the education and employment facilities are also very limited and they commuting to the core city and other area for their livelihood, education and even for marketing. All these problems may also be linked with the deteriorated man-land ratio and underdevelopment which caused heavy congestion and overcrowding of residential establishment. Consequently these areas developed a low class residential area which represents the old city. Many households in this region keep extra families in their houses in rent for their economic profit. Labours, clerks, shop keepers, shop workers, vegetable sellers, auto drivers and porters are seeking residence in this region because the low paying capacity.

The city centre is also overcrowded with market places and business establishments. The working population from outside city are residing in this region for rent due to the vicinity of their work places. This result an insecurity feeling in this region because of the people are not familiar to each other due to their busy schedule and dynamic nature of residences due to shifting of their work place and non permanency in works. But some areas in inner city is kept only for residential purpose like Eranchippalam, Puthiyara, Azhchavattom which not so deteriorated and some areas of which have only elite class residents. The outer or peripheral region can be termed as only residential region which have much open space and generally well maintained roads and drains. The social environment of this region is also very good since a village culture is prevailing in the region with self help groups and residential associations.

The spatial analysis conducted in this study denotes an important connection between the urban physical environment and urban social environment or social wellbeing. Those areas having high physical environmental condition with open spaces, housing structure, infrastructure and facilities like sanitation, water supply, indoor space and the availability of other electronic and information medias are also have better condition in the social factors like small household size, type of family, education, income and security feeling. For example the wards like Silver Hills, Civil Station, Kottoli, Chevayur which have high prosperity in physical environment also have a better condition in the social environmental features like education, family structure and economic viability. On the other hand those areas having deteriorated physical environment in terms of structural and facility perspectives have deteriorated social environment with less security feeling low demographic status and educational and economic backwardness. For example the wards like Chakkumkadavu, Koyavalappu, Pallikkandi have less feasibility in the housing structure especially in open space and per capita indoor space and the facility factors like sanitation and drinking water are also lags in the education, security feeling, family status and the economic prosperity.

The present analysis of the urban environment and social wellbeing in Calicut city suggest that the existing status of urban environment and the social wellbeing is high in comparison with the other Indian cities of same class and population. Even though we can compare the demographic aspects like literacy rate, population growth rate, sex ratio and life expectancy with the cities of developed countries, the infrastructure status like water supply, housing etc. are not up to the mark at least in some parts of the city. The intra-city variations in the urban environment and the social wellbeing are also exists which is not a good sign in the development of a prosperous city environment.

The present study is carried out in mainly at two different scales that is city level and ward level. The smallest unit of analysis used in this study is wards which seem to be not perfect because the intra-ward variations in several features among different *mohallas* of the same wards are easily quantifiable. So if the unit of analysis is taken as *mohallas* the result will be more dependable for the planning purpose and further analysis. It is therefore suggests that any further study should be in *mohalla* wise considering the difficulty of availing data in this level. Another problem, which were faced due to the small number of units of analysis including the selection limited number of variables which could more meaningful with a large number of variables. This was not attempted because of the limitation of the computer programme employed for the analysis i.e. the number of variables must be less than that of the number of observations.

Even though the slum population is very less as compared to the same class Indian cities, Calicut have some slum sites majority of which concentrated in the coastal line. The hygienic conditions in those areas are not fair with congestion of houses, absence of drains, scarcity of clean domestic water and with insufficiency of the building materials. There should be a better housing plan to accommodate them from the side of government and NGOs.

There is a close association between proper environment in terms of shelter and the wellbeing of an individual and family. Inadequate and congenial housing adversely affect both the health and productivity of individual and his family. There are regional disparities in the urban infrastructure and other facilities which is not a good sign for healthy city environment in both physical and social perspectives. The municipal and other government authorities should give their first preference to bring down such huge gap of urban infrastructure and facilities between the low class residential areas of coastal region and the high class residential areas of peripheral region by allocating special funds and innovative housing programmes to the low class region.

Despite of all its limitations, the present study has highlighted the urban physical and social environment of the Calicut city along with the wellbeing and the factors which led to the specific condition prevailing there. But there are certain problems which should be addressed in any future research in the urban environment and social wellbeing of the cities in general and Indian cities in particular. Can the findings of this analysis generalized to the whole medium sized cities or at least to Indian cities? Are there any fundamental differences among cities on which the evolution of the specific characteristics and functions are associated? or if associated can a generalized degree or ratio formulate which can be mathematically represented?

BIBLIOGRAPHY

Andreas J Marchante and Bienvenido Ortega 2006, 'Quality of life and economic convergence across Spanish region 1980-2001', *Regional Studies*, vol. 40.5, pp. 471-483

Andrews, Frank M 1983, 'Population issues and social indicators of wellbeing', *Population and Environment*, vol. 6, no. 4, pp. 210-230

Andrew, Maclaran 1981, 'Area-based positive discrimination and the distribution of wellbeing', *Transactions of the Institute of British Geographers*, New Series, vol. 6, no. 1, pp. 53-67

Anna, Nieboer et al. 2005, 'Dimensions of wellbeing and their measurement: the SPF-IL scale', *Social Indicators Research*, vol. 73, no. 3, pp. 313-353

Astrid, K. Wahl et al 2004, 'Quality of life in the general Norwegian population, measured by the quality of life scale', *Quality of Life Research*, vol. 13, no. 5, pp. 1001-1009

Atiqur, Rahman 1998, *Household environment and health*, B.R. Publishing Corporation, Delhi

Australian Bureau of Statistics 2002, *Social Capital and Social Wellbeing*

Ayyar, K V Krishna, 1938, *The zamorins of Calicut, from the earliest time to A.D.1806*, Calicut

Baleshwar, Takur et al. 2007, *City Society and Planning*, Concept, New Delhi

Barbara, Pisckuri et al. 2002, 'Occupation and wellbeing: a study of some Slovenian people's experiences of engagement in occupation in relation to wellbeing', *Scandinavian Journal of Occupational therapy*, vol. 9, pp. 63-70

Bayless Mark and Bayless Susan 1982, 'Current Quality of Life Indicators: Some Theoretical and Methodological Concerns', *American Journal of Economics and Sociology*, vol. 41, no. 4, pp. 421-437

Benjamin, Colby 1987, 'Wellbeing: a theoretical program', *American Anthropologist*, New Series, vol. 89, no. 4, pp. 879- 895

Beshers, James M 1962, *Urban social structure*, The Free Press of Glencoe, New York

Beteille, A 1991, 'Distributive justice and institutional wellbeing', *Economic and Political Weekly*, vol. 26, no. 11/12, pp. 591- 600

Bilha, Davidson Arad 2005, 'Structural analyses of the quality of life of children at risk', *Social Indicators Research*, vol. 73, no. 3, pp. 409- 429

Bleys, Brent 2007, 'Alternative welfare measure: Over and case study of India', In Kumar Pushpam and Reddy Sudhakara (eds.), *Ecology and Human Wellbeing*, Sage publications, New Delhi

Boal, Frederick W and David N L (eds.) 1989, *The Behavioural Environment*, Rutledge, London

Bonnefoy, Xavier 2007, 'Inadequate housing and health: an overview', *International Journal of Environment and Pollution*, vol. 30, no. 3/4, pp. 411- 429

Bourne, L.S and Simmons J.W (eds) 1978, *Systems of Cities*, Oxford University Press, New York

Brand, Peter and Thomas, M.J 2005, *Urban Environmentalism- Global Change and the Mediation of Local Conflict*, Routledge, London

Braun, Bonnie et al. 2002, 'Welfare to wellbeing transition', *Social Indicators Research*, vol. 60, no. 1/3, pp. 147-154

Breuste, Jürgen 2012, 'The suburban area in an ecological perspective - potential and challenges' In Schenk et al. (eds.) *Suburban areas as cultural landscapes*, vol. 236, pp. 148-166

Brown, A L 2003, 'Increasing the utility of urban environmental quality information', *Landscape and Urban Planning* vol. 65, pp. 85-93

Brown, Scott et al. 2008, 'Built Environment and Physical Functioning in Hispanic Elders: The Role of "Eyes on the Street", *Environmental Health Perspectives*, vol. 116, no. 10, pp. 1300-1307

Brown, Tony et al. 2001, 'Race-Related Correlates of Young Adults' Subjective Wellbeing', *Social Indicators Research*, vol. 53, no. 1, pp. 97-116

Cannon, Richard 2008, *The Social Determinants of Health*, SACOSS (South Australian Council of Social Service) Information Paper

Carbonell, A Ferrer 2005, 'Income and wellbeing: an empirical analysis of the comparison income effect', *Journal of Public Economics* vol. 89, pp. 997-1019

Carlisle, Sandra et al. 2009, 'Wellbeing': A collateral casualty of modernity?' *Social Science & Medicine*, vol. 69, pp. 1556–1560

Carol, L Gohm et al. 1998, 'Culture, parental conflict, parental marital status, and the subjective wellbeing of young adults', *Journal of Marriage and Family*, vol. 60, no. 2, pp. 319-334

Census Reports 1961-2001, Census Commissioner, New Delhi; Ministry of Home Affairs

Chan, Ying Keung et al. 2005, 'Quality of life in Hong Kong: the Cuhk Hong Kong quality of life index', Social Indicators Research, Vol. 71, No. 1/3, pp. 259-289

Chamberlain, Kerry 1988, 'On the Structure of Subjective Wellbeing', *Social Indicators Research*, vol. 20, no. 6, pp. 581-604

Coates, B.E et al. 1976, *Geography and Inequality*, Oxford University press, New York

Coleman, J. S. 1988, 'Social capital in the creation of human capital', *American Journal of Sociology*, vol. 94, pp. 95–120

Cotter, John V and Larry L. Patrick 1981, 'Disease and ethnicity in an urban environment', *Annals of the Association of American Geographers*, vol. 71, no. 1, pp. 40 -49

Daniel, J. Slottje 1991, 'Measuring the quality of life across countries', *The Review of Economics and Statistics*, vol. 73, no. 4, pp. 684-693

Dasgupta, Partha 2001, *Human Wellbeing and the Natural Environment*, Oxford University Press, Delhi

David, M.S. Kimweli and William E. Stilwell 2002, 'Community subjective wellbeing, personality traits and quality of life therapy', *Social Indicators Research*, vol. 60, no. 1/3, pp. 193-225

Davis, Bruce E 1986, 'Quality of life in small island nations in the Indian ocean', *Human Ecology*, vol. 14, no. 4, pp. 453-471

Day, George and Weitz Barton 1977, 'Comparative urban social indicators: Problems and Prospects', *Policy Sciences*, vol. 8, no. 4, pp. 423-435

Desai, Anjana 1986, 'The environmental perception of urban landscape: the case of Ahmadabad', in P.D Mandev (ed), *Urban geography*, Heritattage Publications, New Delhi

Detwyler, Thomas R et al. (eds.) 1972, *Urbanization and Environment*, Duxbury Press California

Diener, Ed et al. 1993, 'The relationship between income and subjective wellbeing: relative or absolute?' *Social Indicators Research*, vol. 28, pp. 195-223

Diener, R.B. 2001, 'Making the best of a bad situation: satisfaction in the slums of Calcutta', Social Indicators Research, Vol. 55, No. 3 (Sep., 2001), pp. 329-352

Dieter, Birnbacher 1999, 'Quality of life: evaluation or description?' *Ethical Theory and Moral Practice*, vol. 2, no. 1, pp. 25-36

District census hand book, Calicut district, Census of India, New Delhi

Douglas, B. Downey et al 1998, 'Sex of parent and children's wellbeing in single-parent households', *Journal of Marriage and Family*, vol. 60, no. 4, pp. 878-893

Drewnowski, J 1971, 'The practical significance of social information', *Annals of the American Academy of Political and Social Science*, vol. 393, pp. 82-91

Drewnowski. J 1974, *On measuring and planning the quality of life*, Mouton, The Hague

Dye, Thomas R 1967, 'Governmental Structure, Urban Environment, and Educational Policy', *Midwest Journal of Political Science*, vol. 11, no. 3, pp. 353-380

Eliezer, Krumbein and Armin Beck 1975, 'The city as a centre of learning, *The Journal of Negro Education*, vol. 44, no. 3, pp. 391-405

European Environment Agency, *The European environment*- State and outlook 2010 Copenhagen, Denmark

Evans, D J and Fletcher M 2000, 'fear of crime: testing alternative hypotheses', *Applied Geography*, vol.20, no. 24, pp. 395-411

Fakhruddin 1991, *Quality of urban life*, Rawat Publication, Jaipur

Felce, David and Perry J 1995, 'Quality of life: its definition and measurement', *Research in Developmental Disabilities*, vol. 16, no. 1, pp. 51-74

Ferriss, A L. 2002, 'Does material wellbeing affect non-material wellbeing?' *Social Indicators Research*, vol. 60, no. 1/3, pp. 275-280

Fischer, Claude S 1981, 'The public and private worlds of city life', *American Sociological Review*, vol. 46, no. 3, pp. 306-316

Fox, Richard G 1977, *Urban Anthropology*, Prentice-Hall, New York.

Ganguli, B.N and D.B Guptha 1976, *Levels of ling in India*, S Chand & Co. New Delhi

Geof, Wood 2007, 'Using security to indicate wellbeing, in developing countries' in Ian Gough and Allister J.M, *Wellbeing in developing countries- from theory to research*, Cambridge University Press, Cambridge

Gibbs 1929, *Travels in Asia and Africa* (tr.) of Ibnu battutta, H A R, London

Gist, N P and Halbert L.A 1954, *Urban Society*, T.Y.C. Co, New York

Goldfield, D R. 1981, 'The urban south: a regional framework', *The American Historical Review*, vol. 86, no. 5, pp. 1009-1034

Government of Kerala 2005, *Kerala sustainable urban development project*- Final Report Volume 2 - City Report Kozhikode,

Graham, Bentham 1986, 'Public satisfaction and social, economic and environmental conditions in the counties of England', *Transactions of the Institute of British Geographers*, New Series, vol. 11, no. 1, pp. 27-36

Groenvold, M et al. 1993, 'Quality of life research methodological issues', *Quality of Life Research*, vol. 2, no. 1, pp. 71-73

Hall, Carolyn 1984, 'Regional inequalities in wellbeing in Costa Rica', *Geographical Review*, vol. 74, no. 1, pp. 48-62

Hammel, Daniel J, 1999 a, 'Re-Establishing the Rent Gap: An Alternative View of Capitalised Land Rent'. *Urban Studies*, vol. 36, pp. 1283-1293.

Hammel, Daniel J 1999 b, 'Gentrification and Land Rent: A Historical View of the Rent Gap in Minneapolis'. *Urban Geography*, vol. 20, pp. 116-145.

Hans, De Witte 1999, 'Job insecurity and psychological wellbeing: review of the literature and exploration of some unresolved issues', *European Journal of Work and Organizational Psychology*, vol. 8, no.2, pp. 155-177

Harrison, G A and J B Gibson (eds.) 1976, *Man in urban environments*, Oxford University press, Oxford.

Harvey, David 1973, *Social justice and cities*, Edward Arnold, London

Harvey Krahn et. al. 1981, 'The quality of family life in a resource community', *Canadian Journal of Sociology*, vol. 6, no. 3, pp. 307-324

Harwood, P L 1976, 'Quality of life: ascriptive and testimonial conceptualization', *Social Indicators Research* vol. 3, pp. 471-496.

Helliwell, John F and Putnam Robert D 2004, 'The social context of wellbeing, philosophical transactions', *Biological Sciences*, vol. 359, no. 1449, pp. 1435-1446

Herbert, D T and Hyde S W 1985, 'Environmental criminology: testing some area hypotheses', *Transactions of the Institute of British Geographers*, New Series, vol. 10, no. 3, pp. 259-274

Herbert, D T and Smith D M (eds.) 1979, *Social problems and the city- geographical perspective*, Oxford University press New York.

Herbert, D T and R J Johnston (eds.) 1984, *Geography and the Urban Environment*, John Wiley and sons, Chichester

Ian, Gough and Allister J.M (eds) 2007, *Wellbeing in developing countries- from theory to research*, Cambridge University Press, Cambridge

Jamilah, Mohamad 1998, 'Building Heaven on Earth: Islamic Values in Urban Development', in Hank Lim and Ranjit Singh (eds.), *Values and Development*, Centre for Advanced Studies, Singapore, pp 53-63.

Jin-Ding, Lin et.al 2010, 'Wellbeing perception of institutional caregivers working for people with disabilities: Use of SHS and SWLS analyses', *Research in Developmental Disabilities*, vol. 31, pp. 1083-1090

Jochem, F.M. and Vander Waals 2000, 'The compact city and the environment: a review', *Journal of Economic and Social geography*, vol. 91, no.2, pp.111-121

Judith, Freidenberg et al. 1988, 'Migrant Careers and wellbeing of women', *International Migration Review*, vol. 22, no. 2, pp. 208-225

Kerala Sustainable Urban Development Project Final Report (May 2005) Volume 2- Kozhikode

Kim, E. and Phyllis Moen 2001, 'Is Retirement Good or Bad for Subjective Wellbeing?' *Jungmeen Current Directions in Psychological Science*, vol. 10, no. 3, pp. 83-86

Knox, P. L 1982, *Urban Social Geography: An Introduction*, Longman, London

Knox, P. L 1975, *Social wellbeing -A Spatial Perspective*, Oxford University press, Oxford

Knox, P L and Cottam M. B 1981, 'A Welfare Approach to Rural Geography: Contrasting Perspectives on the Quality of Highland Life', *Transactions of the Institute of British Geographers*, vol. 6, no. 4, pp. 433-450

Koller, J. M. 1985, *Oriental Philosophies*, 2nd Ed., Scribner, New York

Kunchali, V (ed.) 2004, *Calicut in history*, Calicut university press, Calicut

Lal, A.K 1979, 'Status of Women in a Urban Setting; An Analysis of Role of Differentiation in the Family', *Man in India*, vol. 59, no. 4, pp. 289-297

Lama, Dalai and Cutler H. C 1998, *The art of happiness: a handbook for living*, Riverhead Books, New York

Lapping, Mark B 1979, 'Toward A Social Theory of the Built Environment: Frank Lloyd Wright and Broadacre City', *Environmental Review*, vol. 3, no. 3, pp. 11-23

Leslie, W. Kennedy and N. Mehra 1985, Effects of Social Change on Wellbeing: Boom and Bust in a Western Canadian City, Social Indicators Research, Vol. 17, No. 2, pp. 101-113

Leslie, W. Kennedy et al 1978, 'Subjective Evaluation of Wellbeing: Problems and Prospects', *Social Indicators Research*, vol. 5, no. 4, pp. 457-474

Linda, Bobbitt et al. 2005, 'The Development of a County Level Index of Wellbeing', *Social Indicators Research*, Vol. 73, No. 1, pp. 19-42

Lindstrom, B. and N. Spencer: 1994, *European Textbook of Social Paediatrics* Oxford University Press, Oxford

Liu, Weixin 2000, 'Problems of Environment in Urban areas' in Amitabh Kundu (ed.) *Inequality Mobility and Urbanisation-India and china*, ICSSR New Delhi.

Lloyd, Rodwin and Hollister R.M (eds.) 1984, *Cities of the Mind*, Plenum press, New York

Lydick, E. and R. S. Epstein 1993, 'Interpretation of Quality of Life Changes', *Quality of Life Research*, vol. 2, no. 3, pp. 221-226

Marcia, Grant et al 2004, 'Revision and Psychometric Testing of the City of Hope Quality of Life', *Quality of Life Research*, vol. 13, no. 8, pp. 1445-1457

Maria, Claret et al. 2003, 'Measuring Urban Wellbeing: Race and Gender Matter', *American Journal of Economics and Sociology*, vol. 62, no. 2, pp. 461-483

Marina, Alberti et al. 2003, 'Integrating Humans into Ecology: Opportunities and Challenges for Studying Urban Ecosystems', *Bio Science*, vol. 53, no. 12, pp. 1169-1179

Marne, L et al. 2005, 'The Subjective Wellbeing Construct: A Test of Its Convergent, Discriminant, and Factorial Validity' *Social Indicators Research*, vol. 74, no. 3, pp. 445-476

Matthew, Miller et al 2000, 'Community Firearms, Community Fear', *Epidemiology*, vol. 11, no. 6, pp. 709-714

Mc Gee T.G 1967, *The Southeast Asian City*, Bell and Sons London

Mc Millan S C and M. Mahon 1994, 'Measuring Quality of Life in Hospice Patients Using a Newly Developed Hospice Quality of Life Index', *Quality of Life Research*, vol. 3, no. 6, pp. 437-447

Meier, Richard L 1986, 'Qualities of urban life and evolution of world cities', in Hutchinson and M Battty (eds.) *Advances in Urban System modelling*, Elsevier Science publishers, B.V. North Holland .

Melvin, Pollner 1989, 'Divine Relations, Social Relations, and Wellbeing', *Journal of Health and Social Behavior*, vol. 30, no. 1, pp. 92-104

Michael, Cuthill 2002, 'Coolangatta: A Portrait of Community Wellbeing', *Urban Policy and Research*, vol. 20, no. 2, 187–203

Mikael, Nordenmark 2004, 'Multiple Social Roles and Wellbeing: A Longitudinal Test of the Role Stress Theory and the Role Expansion Theory', *Acta Sociologica*, vol. 47, no. 2, pp. 115-126

Mrozowski, S. A. et al 1989, 'Living on the Boott: Health and Well being in a Boarding-house Population', *World Archaeology*, vol. 21, no. 2, pp. 298-319

Nahid, Osseiran-Waines 1995, 'Social Indicators of Well Being: A Comparitive Study between Students in Bahrain', *Social Indicators Research*, vol. 34, no. 1, pp. 113-152

Nanda and Ravinder Nanda (1977), 'Urban Systems Analysis: An Anthropological Perspective', *Serena Interfaces*, vol. 8, no. 1, pp. 115-119

Narayanan, M.G.S (ed.) 1993, *Malabar* (Malabar Mahotsavam souvenir), Malabar Mahotsavam committee, Calicut

Narayanan, M G S, 2006, *Calicut: the city of truth revisited*, Calicut university press, Calicut

Nathalie, Ostroot and Wayne Snyder 1996, 'The Quality of Life in Historical Perspective France: 1695-1990', *Social Indicators Research*, vol. 38, no. 2, pp. 109-128

Nicholas, Helburn 1982, 'Geography and the Quality of Life', *Annals of the Association of American Geographers*, vol. 72, no. 4, pp. 445-456

Nottridge, H E. 1972, *The Sociology of urban living*, Routledge, London

Ompad, D C et al. 2007, 'Social determinants of the health of urban populations: methodological considerations', *Journal of Urban Health*: Bulletin of the New York Academy of Medicine, vol. 84, no. 1, pp. 142-153

Oropesa, R. S 1995, 'Consumer Possessions, Consumer Passions, and Subjective Wellbeing', *Sociological Forum*, vol. 10, no. 2, pp. 215-244

Osborn, Robert J 1968, 'Crime and the Environment: The New Soviet Debate', *Slavic Review*, vol. 27, no. 3, pp. 395-410

Pacione, Michael 1980, 'Differential Quality of Life in a Metropolitan Village', *Transactions of the Institute of British Geographers*, vol. 5, no. 2, pp. 185-206

Pacione, Michael 1990, *Urban Problems –An Applied Urban Analysis*, Routlege, London

Pacione, Michael 2003a, 'Urban environmental quality and human wellbeing-a social geographical perspective', *Landscape and Urban Planning*, vol. 65, pp. 19–30

Pacione, Michael 2003b, 'Introduction to Urban environmental quality and human wellbeing', *Landscape and Urban Planning* vol. 65, pp. 1-3

Palisi, B J and Canning Claire 1983, 'Urbanism and social psychological wellbeing: a cross-cultural test of three theories', *The Sociological Quarterly*, vol. 24, no. 4, pp. 527-543

Panchayath, level statistics 2006, Kozhikkode, Department of Economics and Statistics, Kerala

Panda, N M and Mishra, B 2001, 'Population Dynamics and Quality of Life in N.E. Region' in Rai et al (eds.) *Environment Resources and Development*, Geographical Society of North-Eastern Hill Region, Shillong

Patsy, Healey 1988, 'The British Planning System and Managing the Urban Environment', *The Town Planning Review*, vol. 59, no. 4, pp. 397-417

Paul, K Hatt and Albert J. Riess (eds.) 1961, *Cities and Society*, The Free Press of Glencoe, New York

Paul, Meadows and Mizruchi E.H 1969, *Urbanism, Urbanization, and Change: Comparitive perspectives*, Addison Wesley Co, Boston

Paul, Amato and Jiping 1992, 'Rural Poverty, Urban Poverty, and Psychological Wellbeing', *The Sociological Quarterly*, vol. 33, no. 2, pp. 229-240

Penny, Warner-Smith and Peter Brown 2002, 'The leisure, health and wellbeing of women in a small, Australian country town', *Leisure Studies*, vol. 21, pp. 39–56

Peri, Kedem-Friedrich and Maged Al-Atawneh 2004, 'Does Modernity Lead to Greater Wellbeing? Bedouin Women Undergoing a Socio-Cultural Transition', *Social Indicators Research*, vol. 67, no. 3, pp. 333-351

Peter, M Fayers and David J. Hand 2002, 'Causal Variables, Indicator Variables and Measurement Scales: An Example from Quality of Life', *Journal of the Royal Statistical Society*. vol. 165, no. 2, pp. 233-261

Philip, R Lee 1967, 'Health and Wellbeing' , *Annals of the American Academy of Political and Social Science*, vol. 373,no. 1, pp. 193-207

Pothna, V et al. 1992, 'Slums of Vishakapatanam; Problems and Policy Perceptions', *Indian journals of regional science*, vol. 24, no.2, 1992, pp. 85-100

Prakash, B A 2002, 'Urban Employment in Kerala; The Case of Kochi City', *Economic and Political weekly*, vol. 37, no. 39, pp. 4073-4078

Putnam, R D 2000, *Bowling Alone: The Collapse and Revival of American Community*. Simon and Schuster, New York

Raphael, Dennis et al. 1996, 'Quality of Life Indicators and Health: Current Status and Emerging Conceptions', *Social Indicators Research*, vol. 39, no. 1, pp. 65-88

Reissman, Leonard 1964, *The Urban Process*, The Free Press of Glencoe, New York.

Rettig, K D and Ronit D L 1999, 'Social A General Theory for Perceptual Indicators of Family Life Quality', *Social Indicators Research*, vol. 47, no. 3, pp. 307-342

Robson, Brian .T 1975, *Urban social areas*, Oxford University Press, Oxford

Rogerson, Robert J 1995, 'Environmental and Health Related Quality of Life: Conceptual and Methodological similarities', *Social Science and Medicine*, vol. 41, no. 10, pp. 1373 1382,

Royuela, Vincente and Manuel Artis 2006, 'Convergence analysis in terms of quality of life in the urban system of Barcelona province', *Regional Studies*, vol. 40.5, pp. 485-492

Ryan, R.M and A.R Sapp, 2007, Basic Psychological needs: A self Determination Theory Perspective on the Promotion of Wellness Across Development and Culture, in Ian Gough and Allister J.M (eds.) 2007, *Wellbeing in developing countries- from theory to research*, Cambridge University Press, Cambridge

Sara, R Curran and Alex de Sherbinin 2004, 'Completing the Picture: The Challenges of Bringing "Consumption" into the Population- Environment Equation', *Population and Environment*, vol. 26, no. 2, pp. 107-131

Sara, S Lanahan et al. 1981, 'Network Structure, Social Support, and Psychological Wellbeing in the Single-Parent Family', *Journal of Marriage and Family*, vol. 43, no. 3, pp. 601-612

Seemin, Musheer and Firoz Khan 2007, *Quality of Urban Environment*, Idarah-i-Adabiyat, Delhi

Sengupta, Ramprasad 2002, 'Human Wellbeing and Sustainable Environment', *Economic and Political weekly*, vol.37, no.42, pp. 4289-4294

Sidaway, J D and Power M, 1995, 'Socio-spatial transformations in the 'post-socialist' periphery: the case of Maputo, Mozambique', *Environment and Planning*, vol. 27 9, pp. 1463-1491

Sinden, J A 1982, 'Application of Quality of Life Indicators to Socioeconomic Problems: An Extension of Liu's Method to Evaluate Policies for 26 Australian Towns' , *American Journal of Economics and Sociology*, vol. 41, no. 4, pp. 401-420

Shir, Ming Shen and S. T. Boris Choy 2005, 'The Pre- and Post-1997 Wellbeing of Hong Kong Residents', *Social Indicators Research*, Vol. 71, No. 1/3, pp. 231-258

Smith, D.M 1973, *The geography of social wellbeing in the United states: an introduction to territorial Social indicators*, McGraw-Hill, New York.

Smith, D M 1975, 'On the concept of welfare', *Area*, vol. 7, no. 1, pp. 33-36

Smith, D M 1977, *Human Geography- A Welfare Approach*, Arnold-Heinemann, London

Smith, D M 1994, *Geography and Social Justice*, Blackwell, Oxford

Srinaivsan, Kartika (2006), Public, Private and Voluntary Agencies in Solid Waste Management: A Study in Chennai City, Economic and Political Weekly, vol. XLI, No.22, June 3, 2006, pp.2259-226

Stanton, E Tuller 1973, 'Microclimatic Variations in a Downtown Urban Environment,, *Geografiska Annaler*. Series A, Physical Geography, vol. 55, No. 3/4, pp. 123- 135

Stewart, Kitty 2000, 'Dimensions of Wellbeing in EU Regions: Do GDP and Unemployment Tell Us All We Need to Know?' *Social Indicators Research*, vol. 73, no. 2, pp. 221-246

Sullivan, Mark D et al. 2000, 'Models of Health-Related Quality of Life in a Population of Community-Dwelling Dutch Elderly', *Quality of Life Research*, vol. 9, no. 7, pp. 801-810

Thompson, Richard A 1973, 'A Theory of Instrumental Social Networks', *Journal of Anthropological Research*, vol. 29, no. 4, pp. 244-265

Tomer, J. F 2001, 'Economic Man vs. Heterodox Men: The Concept of Human Nature in the Schools of Economic Thought', *Journal of Social Economics*, vol. 30, no. 4, pp. 281–293

Tomer, J F 2002, 'Human wellbeing: a new approach based on overall and ordinary functionings', *Review of Social Economy*, vol. 60, no. 1, pp. 23-45

Tore, Saetersdal 1999, 'Symbols of Cultural Identity: A Case Study from Tanzania', *The African Archaeological Review*, vol. 16, no. 2, pp. 121-135

Tran, Thanh Van and Roosevelt W 1986, 'Social Support and Subjective Wellbeing among Vietnamese Refugees',*The Social Service Review*, vol. 60, no. 3, pp. 449-459

Trevor, Hancock 1996, 'Health and Sustainability in the Urban environment', *Environment impact assessment review*, vol.16, pp. 259-277

Turner, R Jay 1981,' Social Support as a Contingency in Psychological Wellbeing', *Journal of Health and Social Behavior*, vol. 22, no. 4, pp. 357-367

Tze, Pin N et al 2005, 'Ethnic Differences in Quality of Life in Adolescents among Chinese, Malay and Indians in Singapore', *Quality of Life Research*, vol. 14, no. 7, pp. 1755-1768

UNO 1954, *Report on international definitions and measurement of Standards of living:* Report by committee of Experts, United Nations Organization, New York

U.S. Department of Health, Education, and Welfare, 1969 *Towards a Social Report*

Valentine, Gill 1989, 'The Geography of women's fear, *Area*, vol. 21, no. 4, pp. 385-390

Vaughan, Laura et al. 2005, 'Space and Exclusion: Does Urban Morphology Play a part in Social Deprivation', *Area*, vol. 37, no.4, pp.402-412

Vinod, Sasidharan 2002,' Understanding Recreation and the Environment within the Context of Culture', *Leisure Sciences*, vol. 24, pp. 1–11

Vlahov, David et al. 2007, 'Urban as a determinant of health', *Journal of Urban Health*: Bulletin of the New York Academy of Medicine, vol. 84, no. 1, pp. 116- 126

Walter, Jamieson 1990, 'Recording the Historic Urban Environment: A New Challenge' *APT Bulletin*, vol. 22, no. 1/2, pp. 12-16

Walter-Busch, E 2000, 'Stability and change of regional quality of life in Switzerland, 1978-1996', *Social Indicators Research*, vol. 50, no. 1, pp. 1-49

Wan, Thomas T. H and Barbara L 1978, 'Interpreting a General Index of Subjective Wellbeing', *Health and Society*, vol. 56, no. 4, pp. 531-556

Weber, A.F 1963, *The Growth of Cities in the Nineteenth Century*, Cornel University Press, New York

Wei Yehua Dennis and Cindy Fan 2000, 'Regional inequalities in China; A case study of Jiangsu province', *The Professional Geographer*, vol.52, no.3, pp.455-469

Weitz, Raanan (ed.) 1973, *Urbanization and Developing Countries*, Praeger Publishers New York

Wendy, D. M and Daniel T.L 1996, 'Parental Cohabitation and Children's Economic Wellbeing', *Journal of Marriage and Family*, vol. 58, no. 4, pp. 998-1010

Wickrama, K. A. S. and Charles L. Mulford 1996, 'Political Democracy, Economic Development, Disarticulation, and Social Wellbeing in Developing Countries', *The Sociological Quarterly*, nol. 37, no. 3, pp. 375-390

William, Michelson 1970, *Man and his urban environment: a sociological approach*, Addison-Wesley Publishing Co, Boston

White, John 2002, 'Education, the market and the nature of personal wellbeing', *British Journal of Educational Studies*, vol. 50, no. 4, pp. 442-456

White, Sarah C et al, 2010. *Religions, Development and Wellbeing in India*, (Working paper- 54), Religions and Development Research Programme

Yongmin, Sun 2001, 'Family environment and adolescents' wellbeing before and after parents' marital disruption: a longitudinal analysis', *Journal of Marriage and Family*, vol. 63, no. 3, pp. 697-713

Yongmin, Sun 2003, 'The wellbeing of adolescents in households with no biological parents', *Journal of Marriage and Family*, vol. 65, no. 4, pp. 894-909

Zenher, Robert 1997, *Indicators of the quality life in new communities*, Balligar Publishing Company, Massachusetts

WEBSITES

http://www.stats.org.uk/factor-analysis/factor-analysis.pdf

http://www.global-greenhouse-warming.com/humanwellbeing.html

http://www.kozhikodecorporation.org/index.php/general-information

http://www.imd.gov.in/section/climate/kozhikode2.htm

http://www.kozhikodecorporation.org/index.php/swm

http://ibnlive.in.com/news/best-cities-to-live-invest-and-earn-in/53060-7.html

http://en.wikipedia.org/wiki/Kozhikode

http://www.kkd.kerala.gov.in/index.php?option=com_content&view=article&id=71&Itemid=53

http://www.kkd.kerala.gov.in

APPENDICES

Appendix: A - Standardized Component Scores

Wards	Factor 1	Factor 2	Factor 3	Factor 4	Factor 5
New Bazar	-3.5522	-14.5205	2.2671	4.2919	0.661
Edakkat	7.5993	5.9547	-3.9757	3.3009	-4.9556
Easthill	8.548	7.7938	-3.7626	1.9649	-8.5897
Kuruvisseri	10.8659	6.0969	-6.946	2.9681	-7.5142
Malaparamba	10.8584	6.4784	-10.7388	-3.8452	-10.258
Vengeri	11.4762	7.3718	-4.4382	3.0257	-5.4691
Kannadikkal	4.4213	4.0539	1.3726	7.0551	-7.5773
Paroppadi	8.7086	2.3082	-3.6518	4.8288	-3.8731
Chevarambalam	8.0494	5.0157	-5.4359	4.4189	-5.1759
Kudilthodu	14.7026	11.2894	-11.5353	1.4334	-7.8258
Cheveyur	15.2277	16.5797	-16.7601	-0.9837	-5.2929
Silverhills	14.0005	10.4238	-14.7404	-0.5503	-6.3624
Poolakkadavu	3.2535	-3.2333	-2.0379	7.5193	-0.703
Moozhikkal	0.271	-3.7786	4.8718	6.8091	-1.0578
Chelavoor	1.3252	-3.0664	8.3341	9.0473	-1.3076
Mayanadu	1.1627	-2.2221	7.6801	7.1057	-0.8949
Kovoor	2.8303	-1.6982	1.6139	3.0943	-2.3733
Nellikkode	10.3742	12.6441	-10.2559	-0.307	-4.1305
Potamal	6.3756	-1.7928	-2.2045	1.7478	-1.2056
Kommeri	3.2851	-0.8443	1.1635	4.1421	-2.7491
Pokkunnu	-2.5465	-3.3038	8.7005	4.3166	2.8819
Mankavu	-0.4593	6.4634	4.337	1.8368	3.1146
Kinasseri	5.9299	-2.8591	-3.2027	3.6659	0.6484
Thiruvannur	12.2428	14.7514	-8.5365	2.7053	3.2483
Kallai	8.0153	-2.8372	-9.6337	-0.7556	-2.0775
Panniyankara	3.9248	-2.4676	-2.855	-0.3517	0.8928
Meenchanda	8.6232	1.9402	-7.8285	5.2522	-1.0216
Koya Valappu	-54.2891	-22.3053	26.3332	-15.3044	15.2026
Payyanakkal	-35.1667	-19.8933	22.0811	-10.3781	13.3227
Chakkumkadavu	-51.3777	-28.5988	37.5019	-8.9081	17.5731
Pallikkandi	-59.4477	-36.3099	31.8753	-9.6747	13.8258
Idiangara	-31.8683	-31.292	10.6548	-5.1404	13.2687
Chalappuram	8.525	8.3694	-3.5543	-0.2822	-0.2454
Azhchavattam	2.5198	9.1438	-0.1986	2.8678	2.0194
Kuthiravattam	10.5951	11.6558	-8.5129	-2.0749	-1.598

Contd.....

Appendix A: Contd...

Wards	Factor 1	Factor 2	Factor 3	Factor 4	Factor 5
Puthiyara	7.9193	7.788	-4.8415	4.421	1.254
Palayam	-4.621	-5.8314	6.2282	-14.4732	-0.265
Big Bazar	-5.0128	-17.1884	2.9493	-15.1361	-0.7194
Vellayil South	-18.4984	-8.4195	19.1169	-8.384	8.4048
Thiruthiyad	5.7849	4.0668	-2.4696	1.3633	0.3788
Kotooli South	12.6667	12.5643	-9.2272	2.9341	-2.7529
Kotooli North	10.3762	9.6711	-7.242	1.5567	-1.9891
Civil Station	16.3105	15.3691	-17.2675	0.1839	-5.7315
Eranchippalam	7.9974	2.2806	-10.1976	-0.8089	-2.0202
Karaparamba	9.0281	6.4278	-7.6821	4.7677	-1.9529
Nadakkavu	10.445	5.5767	-10.1651	1.4196	-0.3159
Vellayil North	-29.0212	-12.9575	15.2559	-9.8276	4.6509
Thoppayil	-16.7307	-8.6764	15.2615	-6.3608	4.6526
Chakkorath Kulam	9.8706	6.841	-7.4786	-0.9281	-0.892
Westhill	12.0589	8.2867	-11.2244	0.507	0.1312
Varaykal	6.4026	6.7965	0.9961	3.9504	2.7908

Appendix B - Selected Variables

S. No.	Variables
	HOUSING STRUCTURE
1	R.C.C. Houses (per 100)
2	R.B.C Houses (per 100)
3	Huts/ thatched Houses (per 100)
4	Owned houses (per 100)
5	Rented Houses (per 100)
6	Houses with open space
7	Houses without open space
8	Houses with 1-2 bedrooms
9	Houses with 3-4 Bedrooms
10	Houses with above 4 bedrooms
	AMENITIES AND FACILITIES
11	House with Sanitation Facilities
12	House without Sanitation Facilities
13	House with Running Water
14	House without Running Water
15	House with telephone/ mobile
16	House with no telephone/mobile
17	House With Electricity
18	House without Electricity
19	Household with Bank Account
20	Household without Bank Account
21	House Subscribing News Paper
22	House with T.V
23	House without T.V
24	House with Computer
25	House without Computer
26	House with Car (per 100)
	DEMOGRAPHY AND ETHINICITY
27	Small households (Up to 4 members)
28	Medium household(5 to 7 members)
29	Large household (More than 7 members)
30	Nuclear Family (per 100)
31	Joint Family (per 100)
32	Hindu population (per 100)

Contd...

AppendixB–Contd...

S. No.	Variables
33	Muslim population (per 100)
34	Christian population(per 100)
35	General Category (per 100)
36	Backward Communities (per 100)
37	Work Participation Rate
	EDUCATION AND INCOME
38	Literacy Rate
39	Family Head with High School Education
40	Family Head with diploma./ Graduation.
41	Low Income Household (Annual income less than Rs. 2.5 lacks.)
42	Medium income Household (Annual income Rs. 2.5- 5 lacks.)
43	High Income Household (Annual income more than Rs. 5 lacks.)
	SECURITY FEELING AND MIGRATION
44	Houses with High Security Feeling
45	Houses with a feeling of Secure
46	Houses with Insecure Feeling
47	Inborn Households
48	Household Migrated within the city
49	Household migrated from outside city

www.ingramcontent.com/pod-product-compliance
Ingram Content Group UK Ltd.
Pitfield, Milton Keynes, MK11 3LW, UK
UKHW021952270726
14060UKWH00002B/477